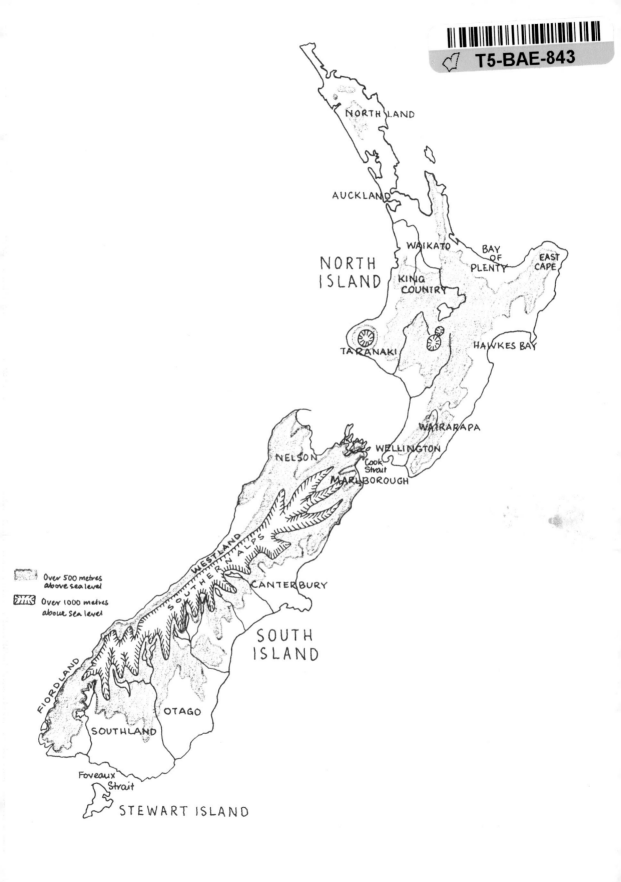

NORTHLAND

AUCKLAND

NORTH
ISLAND

WAIKATO

BAY
OF
PLENTY

EAST
CAPE

KING
COUNTRY

TARANAKI

HAWKES BAY

WAIRARAPA

WELLINGTON

Cook
Strait

NELSON

MARLBOROUGH

WESTLAND

SOUTHERN ALPS

CANTERBURY

SOUTH
ISLAND

FIORDLAND

OTAGO

SOUTHLAND

Foveaux
Strait

STEWART ISLAND

Over 500 metres
above sea level

Over 1000 metres
above sea level

Immigrant
Killers

To Katherine,
with love and thanks

Immigrant Killers

Introduced predators
and the conservation of birds
in New Zealand

CAROLYN KING

Auckland
Oxford University Press
Melbourne   Oxford

Oxford University Press

Oxford London Glasgow New York Toronto
Delhi Bombay Calcutta Madras Karachi
Kuala Lumpur Singapore Hong Kong Tokyo
Nairobi Dar es Salaam Cape Town
Melbourne Auckland
and associates in
Beirut Berlin Ibadan Mexico City Nicosia

First published 1984
© Carolyn King 1984

ISBN 0 19 558121 0

Cover illustration by Chris Gaskin
Designed by Sandra Morris
hotoset in Baskerville and printed by Whitcoulls Ltd., Christchurch
Published by Oxford University Press,
Trentham House, 28 Wakefield Street, Auckland, New Zealand

CONTENTS

Foreword

by Dr Christoph Imboden, Director
International Council for Bird Preservation

A new world opened up for me when I arrived in New Zealand for the first time, just as it had for those early 19th century European explorers, although for different, almost opposite reasons.

The first naturalists landing in New Zealand were stunned by the sight of so many strange birds, many without fear of man and flightless, living in vast impenetrable forests. Ornithologists arriving from Europe today and making their first explorations around the towns and cities of New Zealand have difficulty in believing that they have just traveled halfway around the world. From the familiar voices of the commonest birds, and from the appearance of the landscape and farmland, they could as well be somewhere in Britain. Only when they venture further into the countryside, deeper into the forests and mountains, will they encounter more and more of the bird fauna that had so much caught the fascination of the early settlers and explorers. Sadly, however, some species well known to the pioneers will never be seen by the contemporary bird watcher, because they are extinct, or have become so few in number or have retreated into such remote and inaccessible areas, that only the specialist naturalist will ever catch a glimpse of them in the wild.

I was fortunate enough to work with those specialists and to have the experience of watching and studying some of these vanishing birds in their remote habitats. Their fate in the wake of a giant ecological field experiment induced by the arrival of man in a once pristine land caused my academic interest in ecology and ornithology to change into a feeling of responsibility, a desire to be concerned and to do something about conservation. It was as if I had found a new applied purpose for my years of scientific training. The same, it seems, has happened to Carolyn King: her original interest in the biology of stoats has led to this excellent account of New Zealand's tragic natural history.

The story told by this book has been repeated, in a similar way and with the same devastating consequences, on many other islands the world over. The figures speak for themselves: over 90% of all birds that have become extinct since 1600 lived on islands, and the majority of today's threatened species are also island forms. The particular problems faced by these island dwellers, that had evolved over thousands of years in the absence of mammalian predators and were then suddenly confronted with a whole array of them, are well illustrated here. Although we are now much more aware of them than were our ancestors when they conquered the world's distant corners, the dangers are by no means reduced. On the contrary, man's growing need for economic development, thought to be essential for his own survival, is posing bigger threats than ever to numerous island ecosystems and the survival of countless plant and animal species. There are 930 'high risk' bird species alone, as ICBP discovered in a recent analysis – that is, the 11% of all extant birds that occur only on one single island. Should their habitat on their exclusive home island be destroyed, or be invaded by dangerous predators or superior competitors, they would have nowhere else to go.

As if it was a matter of making up for the fateful, ill-conceived actions of the past, great efforts are now being undertaken to salvage at least part of New Zealand's half-sunken ark. In doing so, the country has acquired a unique expertise in the management of endangered birds. New Zealand scientists are among the world leaders in this field. Through the dedicated and skilful work of officers of the Wildlife Service and other government departments, remarkable successes have been achieved, in the establishment of transfer populations on off-shore islands free of immigrant killers, in the eradication of introduced predators from some important islands, and in the techniques of cross-fostering and double-clutching. Probably the most daring such enterprise is the rescue of the Chatham Island Black Robin, once down to five individuals and, after the 1983/84 season, restored to twenty birds distributed on two islands. Every conceivable technique was employed in this operation, including the planting of a quarter of a million trees in order to create a new habitat on Mangere Island.

The conservation of island bird faunas is one of the principal programme themes of ICBP, with projects being carried out in island groups in all oceans. The active assistance of New Zealand experts has been invaluable for management tasks on the Galapagos, Christmas Island (central Pacific), Mauritius and the Seychelles – all places where introduced mammals are causing severe problems for rare endemic birds and important seabird colonies.

This book is written in such a masterly way that it will be appreciated by the layperson and the expert alike. Above all, however, ICBP hopes that it will be read by people from the many countries that still have unique, irreplacable and often so vulnerable island faunas to look after. We have a collective responsibility for future generations, and we simply cannot allow a repetition of the terrible ecological mismanagement of the past, some of which looks positively criminal today. Further invasions by unwanted alien immigrants must be halted. Ironically, this can be achieved only by the most ferocious of all immigrant killers – by man himself.

Cambridge, May 1984

Note on the International Council for Bird Preservation

The International Council for Bird Preservation, founded in 1922, is the longest-established international conservation organization. It is a federation of over 270 member organizations in 65 countries. Through a vast network of experts ICBP is collecting information about the status of birds and their habitats throughout the world; it identifies priorities for action, initiates and implements projects, disseminates scientific information and advises governments and other bodies on bird conservation matters.

Further information from: ICBP, 219c Huntingdon Road, Cambridge CB3 0DL, England.

The emblem of the ICBP appears on the cover of this book, by permission of the ICBP.

Note on the World Wildlife Fund

The World Wildlife Fund is a non-governmental charitable foundation represented in some 24 countries worldwide and with headquarters in Switzerland.

WWF-New Zealand is the affiliated national organization in this country. It raises funds for scientifically approved projects which will make a practical contribution to the conservation of the natural environment on which all life on this planet – The Earth – ultimately depends.

The panda, emblem of the World Wildlife Fund, appears on the cover of this book, by permission of the WWF-New Zealand.

INTRODUCTION

No one interested in natural history could fail to be riveted by the story of New Zealand. In few other places in the world are such beauty and such tragedy caught up together in so short a span of time and space. So, at least, it seems to me, after 12 years' study in my adopted country.

In 1971, I came to work with the N.Z. Department of Scientific and Industrial Research on the biology of the stoat, a small carnivore first introduced to New Zealand exactly 100 years ago in 1884. Since man[a] first set foot on these islands, the previously undisturbed native wildlife has suffered terrible destruction, to the point that New Zealand and its outlying islands now have the unenviable distinction of contributing 11 per cent of the 318 rare or endangered species listed by the International Council for Bird Preservation for the IUCN Red Data Book – a greater proportion than for any other comparable region.[1] Stoats are known to eat native birds: therefore, research on stoats was then, and is still, officially and publicly expected to concentrate on methods of control, taking it for granted that they *should* be controlled.

Unfortunately, research cannot proceed from assumptions, only from facts. What were the facts concerning the effects of stoats on the distribution and abundance of native birds? At that time the simple answer was: none. In 1984 the answer is still the same. We have more ideas and hypotheses than we had then, but we are still critically short of conclusive facts.[2]

That this dangerous state of ignorance has been allowed to continue for so long is unsatisfactory, but readily explained. One reason is that familiar old bugbear of the natural sciences, lack of finance and manpower.[b] This is an especially savage curb in an isolated country with a relatively small gross income, battling to maintain its accustomed high standards of living in recessionary times. A second reason is that people very commonly underestimate how much time and effort it takes to replace the universal myths and prejudices about a little-known predator such as the stoat with a few simple and indisputable conclusions. Some even confuse the one set of ideas with the other. A third reason is that stoats cannot be considered alone, but as part of a whole suite of introduced predators; and their effects cannot be understood in isolation from all the other massive environmental changes that have nearly overwhelmed the native fauna. This book tries to introduce the general reader to the problems of understanding predators and their effects on native birds, past and present, in New Zealand and elsewhere.

There are about 8600 species of birds alive in the world today.[3] To us they seem as unchanging as the landscape; they have looked more or less the same for as long as there have been people to observe them. But whereas the evolution of birds has spanned the last 165 million years, that of man is much more recent, and covers only a fraction of that time, about three million years. We may find it hard to believe, but in fact the 8600 species of birds that live now are only a small proportion of the number of species that have existed since the evolution of birds began. Estimates of the total number of bird species that have ever lived range from 150,000 to 1,634,000.[4] The difference between these figures shows how difficult it is to make such calculations from the incomplete fossil record, but does not affect the conclusion that thousands of species have appeared, flourished and vanished during the course of avian evolution. Obviously, extinction of old species, replaced by new ones which will be successful for a while and then become extinct and replaced in their turn, is the natural course of evolution. If we are to understand the recent extinctions in New Zealand which have caused so much concern, we must understand the general pattern of extinctions in the past. We must also grasp the essential difference between the infrequent, irreversible total extinction of whole species in evolutionary time, and the frequent, often reversible local extinction of populations in ecological time (p. 30).

One of the basic ideas of modern conservation ethics is that we should protect endangered species, and try to prevent further extinctions, by various means including predator control. But whether we should or could do that depends in part on whether modern extinctions are an extension of a natural, two-stage species-replacement process, or are the result of a much faster chain of events set off by man's own activities, but with the second stage missing. If the extinctions are our responsibility, we should consider what action is necessary and possible: if they are entirely natural, action may not be either necessary or possible. It will help us decide if we examine carefully these questions. Did the bird species that have gone from New Zealand disappear steadily, one after another, or were there some especially bad times? Are all kinds of birds equally liable to extinction, or are some more so than others? What caused the losses, and were they inevitable? Have new species replaced those that have gone, and did they arrive from outside, or develop here? The first four chapters of this book discuss some answers to these questions.

Birds are certainly not the only members of the native fauna to be affected by the changes brought by the coming of man and his assorted companions. But to the average person, losses of native birds are probably more noticeable and more lamentable than losses of, say, native insects, and that was as true a hundred years ago as it is today. So not only do we have more historic records of changes in populations of birds than of insects, but also the problem of protecting what is left of our birdlife touches more hearts than the equally serious problem of conserving native insects. Moreover, birds are sensitive indicators of the health of their environment, and may fairly reflect the state of the less conspicuous members of the natural communities in which they live. This book talks mainly about the conservation of birds, but much of what it says is also valid for other native species, especially the large flightless native insects, the land snails, and the lizards, which are at least as unique and vulnerable to change as are the birds.

As the subtitle shows, this book is about introduced predators. It would be as well to establish from the start what a predator is, and which ones were introduced here.

A predator is an animal that kills in order to eat. Not all animals that eat meat are predators: scavengers and decomposers feed on the carrion of animals they did not kill. (Some both kill and scavenge – a hungry cat or stoat will not turn down a free meal from a fresh road-kill.) But all of us can easily think of examples of predatory animals that kill live prey for food. The most obvious ones are the large carnivores – the lions, tigers, wolves and bears known to every child from picture books and visits to zoos and safari parks. All are members of the mammal order *Carnivora*, the meat-eaters (hence our word 'carnivores').

Lions, tigers and wolves are predators that live up to our expectations. They seldom eat anything but meat, and they usually kill to get it – felling a fleeing deer or antelope and grabbing it with their huge canine teeth. Many of the smaller carnivores are equally fierce and terrifying to animals of their own size.[5] For sheer predatory power and killing ability, gram for gram, few carnivores can match a weasel or a stoat. True, lions and wolves can kill deer larger than themselves, but they hunt in a pack: stoats and weasels tackle rabbits singlehanded.

At the other end of the scale, there are some carnivores which seldom eat fresh meat. Bears fatten up for the winter on fruits and berries; coyotes sometimes cause extensive damage to ripening melons; and badgers eat huge quantities of earthworms – which are of course live prey too, but they somehow fail to excite our sympathies as much as warm-blooded victims do.

Large carnivores such as lions and hyenas are familiar to the public, especially to those who make a point of watching wildlife programmes on television. This is mainly because they are easy to find, exciting to watch, and live in open habitats convenient for filming. But in fact they seldom cause any great changes in their environments, or exterminate any of their prey species, for reasons explained on pages 126-29. The predators that really cause trouble and grey hairs for the world's wildlife service officers are much less conspicuous. The worst of them are not even carnivores, but rodents.

Rats are rodents, and their teeth are adapted for gnawing, not for killing. They do not have the speed and stamina of the dog or the stealth and intelligence of the cat. Nevertheless, they have to be counted as predators, and as very significant ones, because they can destroy enormous numbers of birds' eggs and nestlings. Unlike the large carnivores, which are relatively scarce animals often now restricted to only a small part of their original range, rats can achieve enormous numbers, and have been carried by man in wagons and ships to all corners of the world. When rats invade new territories, especially islands previously free of them, chaos follows.

The predators that came to New Zealand are of four main kinds. The true carnivores are represented by the domestic cat and dog, and three mustelids (members of the family Mustelidae): the stoat, the weasel and the ferret. Anyone asked to name an introduced predator will probably think first of one of these. Then there are the rats – the ship rat, the Norway rat and the kiore or Polynesian rat. A little-known predator is the hedgehog, a member of the Insectivore group which also includes the moles and shrews; it preys mainly on invertebrates such as slugs, snails and worms. And, of course, there is man himself – first the Polynesian race, and then the European. All these predators came from different places, at different times and for different reasons, and they created havoc to different degrees.

The causes of the known historic changes in the birds of New Zealand are still uncertain, although they have been debated for very many years. There is no shortage of suspects, but there are two main problems in sorting them out. One is that two or more aspects of the birds' environments have often changed simultaneously, and the other is that we usually have no reliable evidence collected at the time the most critical changes were happening. In the last century, naturalists could observe the processes of extinction and speculate about the results; today we can observe the results and speculate about the processes. Neither then nor now could all the different factors at work be distinguished and allocated portions of blame. We may never be able to separate exactly the effects of the introduced predators from those of other historic changes, or to decide which of the immigrant killers had the worst effect. But fortunately the predators did not all arrive at once. There was over 100 years between the coming of the first rats and cats, and that of the first stoats and weasels. In Chapters 2, 3 and 4 we review the historical background and geographical setting of each of three more or less distinct periods when different combinations of predators were active, and discuss – so far as is possible in the present state of our knowledge – how far they might have been responsible for the shocking losses of native wildlife over the last 1000 years, in comparison with other changes taking place at the same time. Then we turn to the totally different question of the conservation of our remaining wildlife for the present and for the future.

Many of the facts and ideas expressed here have been presented, in much greater detail but with less background explanation, in various academic journals. References to these papers are listed in the chapter bibliographies, for the benefit of the specialist who has access to primary sources, and who has realized that this account omits a great many of the finer points of the story. But here I am writing for the non-specialist reader, who does not have the previous knowledge assumed in the original papers, or who simply does not want to wade through their technicalities. Details and Latin names that amplify the text, but would interrupt it, are given in the Notes, the Tables (all collected at the end) and the Appendix, although the text itself is complete without them. The reader who is not concerned about verifying any of my statements or pursuing their finer details is thereby enabled to enjoy reading undistracted by academic signposts. On the other hand, I have avoided treating extinctions among native birds, and their relationships with introduced predators, as an emotive subject. Some accounts of it have more of the flavour of disaster journalism than of science, as if non-technical writing had to be arresting in order to be interesting. To me the story is quite dramatic enough already, if it is allowed to speak for itself.

I have written mainly for the many naturalists, bird-watchers and visitors to our National Parks who have helped me over the years and who value the Parks where most of my research was done; also for children, now at school, who will in time inherit the problems this book describes. For them I have included, among the lists of references, easily obtainable and readable sources of further information, such as back issues of *Forest and Bird* and Hamlyn's encyclopedia *New Zealand's Nature Heritage*, and various popular books about New Zealand, predators and conservation in general. I have also had in mind the many people who have rung me to ask for advice on how to catch a stoat: for them, the bibliography gives details of where to find full instructions.[6]

Finally, I have written for scientists concerned with similar problems on other islands, such as Hawaii and the West Indies, who might find here some ideas of interest and a lead into the New Zealand literature.

I regret that some of my conclusions will not please some of the people to whom I owe the most, and that other conclusions are necessarily vague and indecisive. The view I present, while based on scientific evidence, must also be to some extent personal, and other writers may well have treated some parts differently and/or come to different conclusions from the same data. Nevertheless, I have of course been influenced by many discussions with colleagues over the years, and my debt to them, and to the many other people who have helped me in different ways, is gratefully acknowledged below.

C. M. King
Eastbourne, March 1984

ACKNOWLEDGEMENTS

The list of people without whom this book would never have been written is very long. I am grateful to them all, and regret that there is space to name only some of them.

First, H. N. Southern, the long-suffering supervisor of my D.Phil. weasel studies at Oxford; second, J. A. Gibb, Director of Ecology Division of DSIR in Lower Hutt during my employment there. Without these two I might not ever have come to New Zealand in the first place.

During and since my five and a half short years with DSIR, a great many people from the National Parks organization, DSIR, NZ Forest Service and NZ Wildlife Service have supplied me with material and information, or helped me work on it or write it up. Their names are listed in my papers. By far the greatest acknowledgements are due to J.E. Moody and M. G. Efford, for indispensable practical help, and to A. Thorpe, A. Cragg, E. Atkinson and P. R. Dingwall, for equally essential moral and administrative support. I thank all the people who gave me, at various times, help in the field, the laboratory, or the library, and all those who shared data or information, guided me through the minefields of statistics, or criticised my manuscripts. For years of patient typing and retyping of many technical papers, J. Berney deserves a medal. The successive drafts of this book were beautifully typed by B. Cass and B. Huss.

For financial and field support I thank DSIR Ecology Division, N.Z. Forest Service, National Parks Authority, and the Scientific Research Distribution Committee of the Lotteries Board. My debt to the old Fiordland National Park Board is very great and, I hope, obvious from my publications list.

For giving me time and solitude to write, I owe much to the understanding and forbearance of my husband and son, J. W. and D. J. Miller, and to many kind babysitters, especially J. Stuart and B. Cass.

For checking passages of text concerning their areas of expertise, and/or for giving permission to quote unpublished work, I thank R. Pierce, P. R. Millener, M. S. McGlone, B. Lloyd, J. E. C. Flux, C. Ogle, M. Williams, J. Innes, R. Powlesland, P. J. Moors, I. A. Atkinson, G. Kelly and the Wellington and Invercargill offices of the Department of Lands and Survey. The whole text was read by J. A. Mills, C. Daugherty, C. Imboden (for ICBP) and Sir William Gilbert and R. Cleland (for WWF-NZ). Finally, I thank all those who supplied me with illustrations, whose names are given in the captions; C. Cass for her artwork; C. Imboden for writing the foreword; and R. Richards and A. French, who guided the MS through the long processes of production.

There is one other dear friend who would certainly have taken an interest in this book if I had got it started sooner. I have missed not being able to share the joys and sorrows of writing it with Kit Watson, who died on Mount Erebus on 28 November 1979.

The tall ramparts of the Southern Alps rear up behind Three-Mile Lagoon, near Okarito, on the west coast of the South Island. This massive barrier in the path of the prevailing on-shore winds – Mount Cook, centre, now reaches to 3,764 metres – has for two million years brought down four or five metres of rain a year on the lowlands at its feet, supporting luxuriant forest and swamps. N.Z. Department of Lands & Survey.

1

NEW ZEALAND BEFORE MAN
65 MILLION TO 1200 YEARS AGO

The forgotten islands

Once upon a time, let us say somewhere around 1200 years ago, there was a group of green islands huddled together in the southwest corner of the immense blue of the Pacific Ocean.[1] No human tribe had ever occupied them, or given them a name. The two largest islands were very large, and crowned with snowy mountain peaks and smoking volcanoes; spread around them were hundreds of smaller islands, some close inshore, some mere dots on the horizon, some further off even than that.[a]

The fresh cool air brought to them by the West Wind Drift might have been right around the southern end of the globe without touching any land until this, though sometimes it was as cold as if it had just roared in straight off the Antarctic ice itself. Kinder breezes came from the northwest, warmer though not always quieter; the islands sprawled across the Roaring Forties, broadside on to the temperate but boisterous winds of those latitudes. The long chains of mountains down each of the two main islands stood square in the path of the onslaught, like fishing nets in a stream; the rain they caught poured down their western faces in torrents. On the other side of the barrier, the eastern land often stayed thirsty, as the dried-out winds swooped across them and on over the Pacific again.

In the seas around the islands there were places marvellously rich in marine life. Here was the meeting zone of two waters, the main subtropical surface currents of the Pacific and the cold, nutrient-rich deeper flow from the Antarctic. In the turbulence of the meeting the heavier, colder water welled up towards the light, where the nutrients it carried could be used by planktonic algae; these in their turn supported tiny planktonic hunters, then krill, then vast schools of fish. These rich hunting grounds were the home of uncounted thousands of seabirds, seals, sealions, dolphins and whales, which congregated on or near the shores of every island in the group to rest, moult and breed. In spring when the seabirds came in to nest, the noise and the crowding of the breeding colonies must have been intense – and they were not confined to the coasts, as some of them nested in the mountains far inland. Large groups of whales cruised slowly in the shallows, blowing and splashing with their tails, while the dolphins winnowed gracefully in and out of the waves. Seals and sealions jostled for position on the crowded beaches.

All the islands were covered from head to foot, hilltop to coast, with thick evergreen forest. On the driest eastern hills, on windswept ridges, in frosty alpine valleys and on the coldest southern islands, the forest thinned out, degenerated into scrub or even disappeared; and of course it could not grow above the natural tree line at about 1000 metres up the highest mountains, or on sand dunes or shingle banks. In the dry hills of central Otago, where droughts and lightning fires were relatively frequent and winters harsh, the tallest vegetation was open woodland interspersed

The Rakaia River flows eastward from the Southern Alps, its bed choked with gravel and debris carried down from the mountains. The west coast is often cloudy or wet whilst the eastern plains bake in the sun: the original eastern forest was drier, sparser, lower, and much more vulnerable to fire than the dripping western bush. Author.

The Oparara Valley in Northwest Nelson State Forest Park is still covered with trees for as far as the eye can see. Much of New Zealand looked like this 1200 years ago, when the first human colonists arrived. S. A. Bartle, National Museum.

with scrub and grassland. But otherwise, the forest was supreme – a solid green
blanket, over 21 million hectares of it altogether (Table 1).

In the warm, wet lowlands the forest was many-layered and thick enough to shut
out almost all sunlight from reaching the ground.[2] Immensely tall trees towered up
toward the sun, spreading their first branches only when the trunks were eight or ten
metres clear of the soil. Each one of these forest giants was at least several hundred
years old, and surrounded by a host of smaller broadleaved trees, shrubs, saplings,
and seedlings. In damper spots grew the graceful treeferns, spreading their feathery
crowns from the tops of shaggy trunks. On the forest floor the leaflitter lay thick and
damp in semi-darkness. Ground ferns and fungi of every size and shape thrived in
such conditions; clematis and orchids glowed brilliant white in the gloom, but even in
the height of summer the predominant scent was of leaf-mould. Myriads of winged
insects flashed through shafts of sunlight; huge wetas, cricket-like flightless insects,
stalked or jumped through the carpets of fallen leaves; tiny brown and green frogs
and smooth brown skinks scuttled about their business, and enormous snails with
massive pointed or whorled shells inched ponderously forward on theirs. But the
lords of the forest were the birds.

The air was alive with the sight and the sound of birds, calling from the green leafy
ceiling overhead, hopping through the canopy and down the tree trunks, turning
over the leaflitter and poking into every hole, oblivious to any thought of looking out
for enemies. Nor was there any need, for few birds of prey waited to pounce from
above, and no ground hunters – no mammals, no snakes – had ever prowled the
forest floor. The birds and other native animals of those times may not have lived in
perfect peace and harmony, but at least the causes of whatever disturbances they
suffered never included violent death in the jaws of any reptile, carnivore or human
hunter. The forests of New Zealand, and even some of their inhabitants, were pretty
much as they had been around 130 million years ago, long before the modern
carnivores had evolved, and aeons before man's remotest ancestors were even a
twinkle in the eye of the Creator.

How is it that, long after the Roman Empire had risen to and fallen from
unprecedented heights in human civilization, life in New Zealand was still much as it
had been millions of years ago in the pre-human past? How could a whole country of
that size remain undisturbed and unchanged for such a long time, and until so
recently?

The answer to this question – and to many other questions which geologists have
been asking for years – is best answered by the modern theory of plate tectonics or
Continental Drift.[3] Many kinds of studies, in widely differing fields of science, have
shown that the surface of the earth is divided into huge sections or plates, lying like a
cracked skin on a layer of semi-fluid or 'plastic' rock between 60 and 150 kilometres
down – that is, far below the roots of the mountains or the beds of the deepest oceans.
The plates are of varying size, and all are moving relative to each other, carrying the
continents and oceans as passengers on their backs. Over the last 150 million years
the continents have been slowly moved about on the surface of the earth, drastically
changing their positions relative to the latitudes (i.e. to the warmer and colder
climatic zones) and also relative to each other. And of course all the plants and
animals living on the moving pieces of jigsaw have had to go along too.

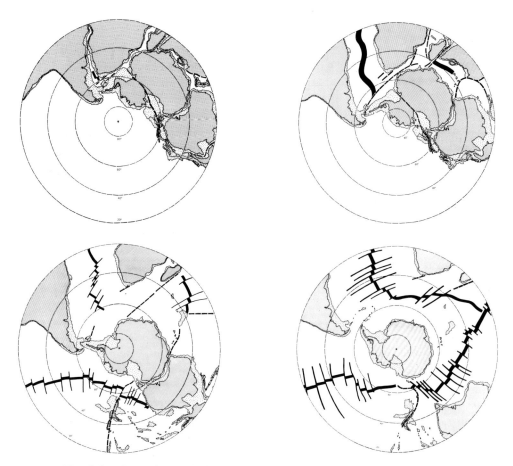

(Top left) *The southern end of the world at about 120 million years ago, in the early Cretaceous. The dotted lines show rift zones (cracks or faults in the earth's crust) and the thick bars show the zones under the oceans along which the crustal plates were being forced apart. For the previous 60 million years, all the southern lands were joined together in a single mass, called Gondwanaland, over which the plants and animals living at the time could disperse in any direction. New Zealand's most ancient species, including moa, kiwi and the dominant forest trees, were probably already in residence when the southern lands began to split, starting in the South Atlantic, at about this time.*

(Top right) *The southern end of the world at about 90 million years ago, in the late Cretaceous. Antarctica, Australia and New Zealand are still joined together, but are drifting towards the cooler polar latitudes and away from South Africa, Madagascar and India – that is, from all direct contact with the animals and plants evolving in Eurasia and Africa.*

(Bottom left) *The southern end of the world at about 60 million years ago, in the Palaeocene. The snakes and marsupials arrived in Australia about this time, but New Zealand had already drifted off into complete isolation.*

(Bottom right) *The southern end of the world today. The separation of Antarctica from South America and from Australia has opened the way for the circum-Antarctic current and the West Wind Drift, which endlessly circle the southern end of the globe, and carry many colonising animals and plants to New Zealand. G. Stevens.*

The story of New Zealand starts in the late Cretaceous period, about 65-90 million years ago. Until then, the fragment of continental crust which later became twisted and displaced into its modern position as the New Zealand archipelago lay close alongside a much larger mass comprising the present Australia and Antarctica joined together. This larger mass was itself part of an even huger one, the ancient super-continent of Gondwanaland.[b]

Around that time, the climate of the infant New Zealand was never cooler than it is now, and was sometimes subtropical. The forests contained many of the trees and shrubs that still live here, mostly the somewhat primitive types that have since disappeared elsewhere in the world. The modern flowering plants, or angiosperms, became widespread only about 100 million years ago, and few of them reached New Zealand before the beginning of its isolation.

We cannot be certain what animals lived in these forests, since animal fossils are rare compared with those of plants, and the chances of preservation of land animals are remote. But we know that about 130 million years ago, before the split, the ancestral New Zealand landmass had, probably for the only time in its history, land links with all the Southern Continents: north-west to the Indo-Pacific and Southeast Asia, south-east to Antarctica and, on the other side of it, South America. It seems reasonable to suppose that the land animals that lived then in what was to become New Zealand were representatives of the kinds of animals living on those continents at that time. They did not include any marsupials or snakes, because by the time these animals arrived in Australia, about 60 million years ago, New Zealand was already isolated.

So as New Zealand, an isolated fragment split off from the side of the Antarctica/Australia landmass, drifted further and further eastwards as the Tasman Sea widened, it became what Tim Halliday aptly describes as a 'half-filled Noah's Ark, which took with it only a few of the species living at that time, and which left too early to take many of the animals and plants that have since become a major part of the world's fauna and flora'.[6]

For most of the next 65 million years (the Cenozoic era) the land of New Zealand looked nothing like it does now; it was generally low-lying, an archipelago of islands changing in shape from time to time as the processes of erosion and deposition joined or separated them across shallow seas. The west winds brought seeds of many new plants to add to the ancient forests, while those already resident evolved into different forms filling different niches in the various forest communities. The Tasman Sea, now 1900 kilometres wide, was not an effective barrier to the flying birds of Australia: many of them have dispersed across it since it was formed, as they still do in the present day.

About seven million years ago, the earth movements which had begun around 20 million years ago climaxed in the tremendous upheavals which built up the present Southern Alps and North Island volcanic mountains. The crustal plate bearing the western Pacific Ocean was being driven by unimaginable forces into head-on collision with the plate bearing the Tasman Sea, Australia and part of the southern Indian Ocean. The edge of the Pacific plate was being slowly ground under; surface rocks, mud, fossils and all were forced down into the molten core of the earth, not smoothly, but juddering and sticking all the way; the crust shook with earthquakes

The St Arnaud Range, in Nelson Lakes National Park, from Mt Robert. Beech forest clothes the sides of a valley so steep that Lake Rotoiti, filling the bottom of it, is quite invisible. Subalpine grassland stretches from the treeline at about 1400 metres to the top of the range, 400 metres higher. Author.

and split with cracks or faults. In the enormous heat and pressure down below, the dragged-down surface material melted into the fiery magma of the earth's interior. The Southern Alps were formed (and are still being formed) from the buckled edges of the two plates, jerked upwards, often many feet at a time, whenever a large earthquake temporarily relieved the strains; and wherever a fissure in the weakened crust allowed, outpourings of molten magma piled up the huge volcanoes of the North Island.

On the surface, most of the forests lived on unperturbed by the geological agonies going on under their roots. Some were swept away by massive landslides triggered by earthquakes, or drowned in lakes created when such landslides fell across a river. Sometimes large areas of forest were smothered by ash from one of the more spectacular volcanic eruptions. But such disasters never covered the whole country, so there was always much more forest left than was destroyed, providing new animal and plant settlers to reoccupy the devastated areas. The mountain-building processes affected the resident fauna and flora only to the extent of providing a new habitat, the alpine zone. Here was a new theatre for the evolution of alpine forms of grasses and daisies, parrots and wrens, as soon as the young mountains reached above the altitudinal limit for trees. Further down, however, life went on more or less as before. It is strange to think that all the oldest of New Zealand's resident species of forest animals and plants have been here longer than the mountains they now live on.

About two million years ago a series of far more devastating environmental upheavals began, the violent climatic fluctuations of the Pleistocene period, better known as the Ice Ages. That name is rather misleading, because the continental glaciers were very unstable, advancing and retreating many times; in the intervening periods, the interglacials, the climate was at least as warm as now. However, the glacial stages were certainly spectacular enough to command the most attention. During the latest one, the sea level dropped by at least 100 metres, uniting many previously separated islands in the New Zealand archipelago into one; permanent snow reached below 350 metres altitude, eliminating the forest everywhere except in Northland, the northern coasts and the exposed seabed between the islands; and the average annual temperature dropped by between 4.5 and 6°C.[7]

The native fauna that had been resident during New Zealand's 50 or 60 million years of quiet and temperate isolation must have been devastated by these changes. In Europe and North America, at least some warmth-loving species could migrate southwards and sit out the cold glacial stages in the Mediterranean and Central America; but there was no adjacent warmer land free of ice to which New Zealand animals and plants could escape.[c] The only refuge at home was a relatively small area of surviving warm-temperate forest in the north of the North Island, and the more extensive areas of cool temperate forest further south. The Pleistocene ice ages probably wiped out many more native species than the few we know of from fossils, no doubt including some of those which arrived, during each of the warm interglacial periods, to take up the spaces left vacant by their predecessors. When the effects of the most recent glacial stage finally wore off, the forests slowly re-established themselves between 14,000 and 10,000 years ago and new trans-Tasman immigrants began the replacement process all over again.[8]

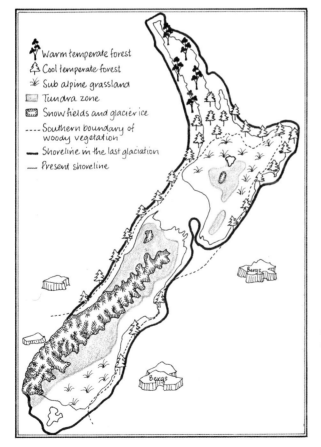

During the glacial phases of the Ice Ages, New Zealand was much larger than now, because huge areas of seabed were exposed by the 100-metre fall in sea level. Permanent snowfields and glacier ice covered the highest mountains, and in the southwest of the South Island ice stretched out to sea. Warm temperate forest survived only in Northland; all other ground where trees could survive at all (including the exposed seabed) was occupied by cool-temperate beech forest, kept at a respectful distance from the ice by broad intermediary zones of subalpine grassland and tundra. Alpine and grassland species of native and immigrant plants and animals thrived; warmth-loving forest species huddled in Northland, or died out. Cynthia Cass.

Warm-temperate forest once covered enormous areas of the North Island, with huge native conifers (podocarps) thrusting olive-green crowns over the heads of the smaller trees and treeferns below. Together they cut out so much light that the animals and plants of the forest floor lived in semidarkness, but the immense variety of trees, shrubs and smaller plants made the podocarp forests a much richer source of food than the beech forests. Guy Salmon.

The native forests and wildlife

The forests of those times were immensely rich and diverse. There were two main kinds, one dominated by the southern beeches, and the other by native conifers. They were both evergreen, and so never blazed with autumn colours like the deciduous forests of the northern hemisphere, but they made up for that in their special character.[9] The forests of New Zealand still represented, after millions of years, the only untouched and more-or-less unchanged survivors of the forests of ancient Gondwanaland. Even today, graceful waving treeferns grow here that would not look out of place in a museum diorama filled with dinosaurs.

The various kinds of warm-temperate lowland forests covered huge areas of both main islands and the whole of Stewart Island. In the north of the North Island, many subtropical plants and trees from New Zealand's ancient, warmer past did manage to survive the glacial periods, and then to re-establish themselves after the end of the latest one. Huge emergent podocarps (native conifers) and rata (a brilliant flowering tree related to the myrtles) stood tall above a canopy of larger hardwood trees, and below them there was a subcanopy formed by an immense variety of smaller trees. The largest of the warmth-loving species, the majestic kauri, towered up to 35 metres high on a massive straight, untapering, stone-coloured trunk, and spread its grey-green foliage clear over the heads of all other trees of the forest. Creepers and vines scrambled upwards from dark, impenetrable thickets on the ground, and huge clumps of epiphytes, or perching plants, made use of the height and strength of their neighbours to get a 'leg-up' to the light. Further south the kauri and the other strictly subtropical species dropped out; but the southern podocarp forests were still immensely rich and diverse, with many different kinds of plants and trees producing fruits, berries or nectar in a generous sequence all the year around.

The cool-temperate forests of southern beech formed continuous, solid blankets, softening the angular outlines of the young mountains of the North Island, and lapping like a soft green sea up the sides of the taller Southern Alps of the South Island. In places the beech forest reached down to the sea, and in other places it was absent altogether, even at high altitude, so it is not entirely true to say that cool, mountain forests were always of beech, but they usually were. Wherever they could find deep soils and shelter from the wind, red beeches could grow into tall, straight timber trees; at the other extreme, the mountain beeches forming the treeline were often twisted and stunted by the cold and shorn by freezing gales.

Beech forests were generally quite simple in construction: a solid, interlocking, single-layered canopy of small, evergreen leaves, thick and shiny to resist cold, and down below, a thick springy carpet of leaflitter, moss and ferns, criss-crossed with rotting logs. The further up the mountain-sides these forests grew, the less luxuriant, lower in stature and poorer in species they became, as the trees characteristic of the warmer lowland forests all reached their upper limit long before beech. So the number of other kinds of trees and shrubs mixed in with the beech thinned out with altitude. Every three or four years, when the beech trees had a 'mast' year, they produced tons of small, hard triangular seeds and dumped them *en masse* on the forest floor. In other years the beeches produced little or no seed, and then the most abundant alternative foods for forest animals were the myriads of small creatures in the leaflitter, and the millions of flying and crawling insects.

These two kinds of forest were very different in appearance, inhabitants and distribution, but they mingled in various ways, and, particularly in the lowlands, they teemed with wildlife, and especially with birds. Both the forests and their inhabitants were quite unlike any forest ecosystems found anywhere else in the world, for two

Cool temperate forest, dominated by the southern beeches (Nothofagus) *grew in the part of Gondwanaland that later became New Zealand, and also in the other parts of it that later became eastern Australia, Tasmania, Antarctica, Chile and New Guinea.* Nothofagus *seeds are very poorly equipped for long-distance travel across the sea, and so are certain small primitive animals that are only ever found in southern beech forests. These patterns of distribution are often taken as among the strongest biological data supporting the theory of continental drift.* Author.

main reasons. The first is that remnants of the ancient flora and fauna of Gondwanaland had been protected here by prolonged and remote isolation, while their relatives elsewhere in the world were being displaced by more advanced forms that evolved later. The second is the total absence, for all that time, of any mammalian or reptilian ground predators.

The real aristocracy of New Zealand's original wildlife were the archaic species, those that were already in residence when the land they lived on was still part of Gondwanaland. Having found a workable way of life that was not drastically changed by any outside interference, many of them saw no reason to change either. Among the very oldest residents was *Peripatus*, a zoological oddity about five centimetres long and looking like a green velvet caterpillar with legs. It was still pretty much the same as it had been since the Cambrian period, around 550 million years ago. The tiny native frog, *Leiopelma*, still resembled the primitive frogs found as fossils in Jurassic rocks around 150 million years ago, and still had rather unfroglike squeaks as well as muscles to wag the tail its ancestor had not long discarded. The tuatara, a lizard-like reptile with large eyes and a crest of short spines down its back, was not a lizard at all, but the sole remaining member of a separate order of reptiles that first appeared about 220 million years ago. All the once-abundant relatives of the tuatara elsewhere in the world became extinct with the dinosaurs, about 65 million years ago: it survived alone on the forgotten islands of New Zealand, still practically identical in structure to its close cousins of 140 million years ago. Among the smaller creatures of the forest floor, there were earthworms, insects, snails, and slugs with equally ancient lineages.[10]

The animals that make up the marvellously diverse faunas of the major continents are constantly under pressure from a great variety of competitors and predators. The only animals that survive are those which manage to keep one jump ahead in some way – perhaps to feed more efficiently, build more defences, keep the sharpest lookout or whatever other strategy works best. But all these ploys take time and energy, and they also close off other options for increasing efficiency. For example, flight is an enormous advantage to help a small bird or insect to escape from predators, but it imposes equally enormous costs, in the energy required for flying and in the limit that flight imposes on body size. If the need for rapid escape is removed, that energy could be better invested in building a larger, less mobile, but more efficient body. One whole family of archaic New Zealand insects took this course, the giant flightless cricket-like wetas. But by far the greatest benefits of

Giant flightless wetas were common in the mainland forests of primeval times. In the absence of any terrestrial mammals, either to compete with them or to eat them, the wetas developed habits and lifestyles similar to forest rodents elsewhere in the world – they holed up during the day, and emerged at night to feed on forest floor debris. These 'invertebrate mice' even produced rodent-like droppings! N.Z. Wildlife Service.

abandoning flight and increasing body size are reaped by warm-blooded animals, especially birds. The energy costs of flying are equalled only by those of keeping a small body warm in a cool climate. Small wonder that an unusually high proportion of New Zealand's ancient native birds tended to become large and flightless.[d] In P. R. Millener's survey of the subfossil birds of pre-human Northland, 47 per cent of the 53 species represented in one study area were flightless forms.[11]

The most spectacular of them all were the moa, which had taken the advantages of large size to their logical extreme. With their smaller cousins the kiwi, the moa were all so completely flightless that they showed no evidence of ever having been able to fly. For example, they had no 'keel' on the breastbone, which in other birds is the anchor-point for the large flight muscles. There were about a dozen species of moa, ranging from large (turkey-size) to gigantic, and three species of kiwi, all much smaller (domestic hen-size). The moa and kiwi were closely related to each other and different from their nearest relatives (ostriches, emus and cassowaries), so they must have been isolated together for a long time. At least the smaller species were well adapted to living in forest – indeed, there has scarcely been any other kind of habitat in New Zealand for most of its history.[e]

The moa diversified in height and therefore, presumably, also in habits, and were found in great abundance over both main islands. The largest, the giant moa or *Dinornis*, is usually said to have stood three metres tall when its long neck was erect; in

The ancestors of the moa and kiwi were living in New Zealand before it was isolated from the rest of Gondwanaland, and they continued to live in the swamps and forests of a generally warm and low-lying island or group of islands for the next 40-45 million years. They survived the slow cooling of the climate and the building-up of mountains under their feet during another 18 million years; they even survived the relatively more rapid climatic oscillations of the last two million years. Given enough time, space and freedom from interference, animal populations can sometimes be capable of remarkable resilience. Cynthia Cass.

that position – the one in which all the early skeletons were mounted – it would have
overtopped any living mammal except the giraffe or the elephant. Modern studies
suggest, however, that the neck was usually held looped forward rather than straight
up, so that a moa's highest point was the middle of its back, making it probably less
than two metres tall.[12] It had a small head, a huge rounded body covered in loose
shaggy feathers, and legs more massive than those of a draught horse, covered in
reptilian scales. Even the smallest moa probably stood a metre high, and had the
same general profile. Fossilized moa gizzards have been found to contain the twigs
and leaves, fruits and seeds of the common vines, shrubs and trees of the forest,[13]
which, since moa had no teeth, had to be ground up by stones in the gizzard. Their
eggs were huge, nearly as big as a man's skull, and probably laid in small (compared
with ostriches and emus) clutches of about four.

The kiwi were much like moa in their shape, their shaggy feathers, and their large
strong feet, and also in having enormous eggs, laid only one or two in a clutch. But
instead of short triangular beaks for browsing by day, the kiwi developed slender,
elongated beaks for probing in soft forest litter for worms, insects and grubs at night;
and instead of being one to three metres tall, the adult kiwi stood a mere 35
centimetres, probably not much bigger than newly hatched moa chicks. They
emerged from their burrows at dusk, calling and whistling to each other, and
foraged through the litter snuffling through their nostrils, which were placed at the
tips of their long beaks. Their eyes were weak, and in any event not much use in the
forest at night, so they had to rely on a sharper sense of smell than most birds, and on
sensitive bristles on the face, to find their way in the dark. The female was larger than
the male, and left the prolonged incubation of their egg (about two and a half
months) entirely to him.

The moa and kiwi shared the forest floor with a great variety of other birds, some
of which might have been with them in Gondwanaland. Most of the rest had flown
across the widening Tasman Sea at various times since the split, usually by
hitch-hiking from Australia on the West Wind Drift. Finding rich and extensive
forests free from any of the terrestrial predators of their homeland (snakes, rats,
dingoes and carnivorous marsupials) many of these new arrivals quickly saw the
advantage of reducing their powers of flight, or even abandoning them altogether.
The extent to which the colonizing birds adapted to life in New Zealand, and became
recognizably different from their ancestors, shows how long they have lived here.
The processes of evolutionary adaptation work so fast that only the most recent
arrivals (those coming within the last few hundred years) are still identical with their
parent stocks overseas.[8]

The scene of 1200 years ago was lively with all kinds of birds, in an abundance and
diversity that we who are familiar with the New Zealand forests of today can scarcely
imagine. Among those species which had probably been in residence for a very long
time were several other large flightless forms besides the moa and kiwi. There was a
giant rail, *Aptornis*, which stood a metre tall – comparable with the smaller moas – but
with a much larger, stronger beak. There was the takahe, a smaller (63 centimetre)
but stoutly built bird clothed with iridescent plumage of blues and greens, and a
bright red beak and frontal shield. There were also several other kinds of rails, a
great variety of waterfowl including ducks, coots, swans, pelicans, herons, snipe,

Harpagornis was probably one of the largest eagles ever known in the world. It had relatively longer legs and shorter, broader wings than other large eagles, but its wingspan was still two to three metres across. It sometimes fed on the bodies of moa mired in swamps, and – since it weighed at least 10 to 12 kilograms – it occasionally also became mired itself. This artist's impression is based on a reconstruction by Canterbury Museum archaeologists.[12] Cynthia Cass.

metre-tall flightless geese (*Cnemiornis*), a harrier hawk, the fearsome *Harpagornis*, and another distinct kind of fish-eagle, which lived in the Chatham Islands.

The forests too were full of unique birds of very ancient lineage. The huge, fat New Zealand pigeon, with its green head and snowy chest, fed on the berries of forest trees and shrubs, and was an important agent of dispersal for the podocarp and hardwood forest trees. Groups of kaka (a large, noisy parrot) squawked and quarrelled among the tree tops. Flocks of small green parakeets sporting bright red or yellow caps congregated on bushes laden with berries. In the sunny upper canopy, three kinds of honeyeaters probed for nectar with their long bushy tongues, pollinating the flowers in the process: the bellbird and tui especially were magnificent songsters, and made the forest ring with a riotous chiming of melodious bell-like notes[f] punctuated by a variety of unexpected and hilarious grunts, croaks, clicks, coughs, rattles, wheezes and chuckles. On the trunks and branches of the trees, at least five kinds of minute, almost tailless green or brown wrens ran up and down, searching for insects in the bark.

On the ground, a large long-legged, slaty-coloured robin (not a robin at all, but an Australian flycatcher with a robin-like shape) waited, watching intently with large bright black eyes, for any small creature to stir among the leaves. Equally intent on the same task, grubbing among the dry leaves and forest debris, was a group of piopio, or native thrushes. The flightless weka, or woodhen, stalked through the undergrowth pouncing on unsuspecting frogs and insects with its sharp, stabbing bill. Pairs of huia, resplendent in black plumage with white-barred tails, fed close together, the males breaking up rotten wood with their sharp, straight stout beaks, and the females reaching into crevices to pull out grubs with their long, slender, curved beaks. Great numbers of saddlebacks and kokako searched for insects and

*Most parrots are tropical or subtropical forest birds: the kea lives in the cold,
windswept, treeless alpine zone of the South Island. As the Southern Alps inched
upwards with each jerk of the Alpine Fault, the ancestors of the kea moved up with
them, and adapted to the new habitat as it was being created. Their close relatives, the
kaka (p.137), stayed in the lowland forests.* Author.

fruit on the forest floor or in the lower canopy, using their long legs for running or hopping, and their wings only for gliding or for very short flights. At night, the male kakapo, a flightless, mossy-green parrot, a larger version of the night parrot of Australia, constructed intricate systems of runways and display areas, and in spring spent hours at a time 'booming' to attract a mate. The noise was increased when the laughing owl, a large long-legged hunter of wetas and lizards on the forest floor, gave warning of its approach with repeated, loud and dismal shrieks, followed by a series of whistling, chuckling and mewing notes. Large crows pecked systematically at rotting carcasses, performing the useful tidying-up job that carrion feeders do everywhere. All these birds had lived for countless generations in the rich native forest, and all were completely dependent on its existence for their survival – that is, for food, shelter and a stable, predictable environment. Even the curious blue duck, which searched for water insects under stones in the beds of fast streams pouring off the hills, depended on the forest to absorb sudden downpours and keep the torrents flowing clear and steady. All of them were endemic – unique to New Zealand, and found nowhere else in the world.

The drastic environmental changes caused by the Ice Ages must have eliminated many of the warmth-loving endemic birds; but equally they created new habitats which were in time colonized by new arrivals. By 1200 years ago these were as abundant and widespread as the older residents, but their recent arrival was betrayed by their stronger similarity to their relatives overseas. The characteristics they still retained included the ability to cope, to various degrees, with the attentions of the predators common in their homelands. As might be expected, few were strictly forest birds. The crested grebe, white-throated shag, black-fronted tern, brown duck, shoveler, marsh crake, pipit, quail, dotterel, black-billed gull, paradise duck, scaup and merganser all revelled in the extensive open country and abundance of fresh water typical of the glacial and post-glacial periods. Some birds already resident took advantage of the changes and became adapted to cold-climate habitats, for example, the kea (a large green parrot) and the rock wren (a brown relative of the bush wren). Other colonists arriving about or since this time – the fantail, banded rail, kingfisher, warbler and tit – were happy to live in forest when it reappeared, but did not need it in the way the older residents did.

The kakapo is the world's heaviest parrot, and the only flightless one. It has an owl-like face with a broad, strong beak framed by 'whiskers', and it strides through thick forest and undergrowth at night, occasionally jumping or gliding on its short, round wings.
Cynthia Cass

As a group, the seabirds of New Zealand have a rather longer history than the land birds. About 30 million years ago the drifting continents parted far enough to open Drake Passage, the sea gap between South America and the projecting tip of Antarctica. From then on there was no land in the way of the circumpolar winds – the West Wind Drift – and the soaring seabirds that rode on it. The New Zealand islands offered safe, predator-free nesting sites within reach of the rich marine resources of the subantarctic, and so there have been dense seabird nesting colonies here for at least 20 million years. As the *New Guide to the Birds of New Zealand*[15] says, the true glory of New Zealand birds is best seen and appreciated in her coastal waters. Most of the great variety of marine species now resident have been here much longer than any but a handful of the land birds.

The risks and benefits of life on islands

The evolution of the native fauna of New Zealand illustrates the basic truth that wild nature is dynamic, changing all the time. The impressive mountains and forests look permanent, because natural changes in them are slow compared with the length of the average human lifespan. But in geological terms the present landscape, and quite a lot of its contents, are brand-new. Many birds that once thrived here have gone, and others which are common now were unknown a hundred years ago. The story of New Zealand bird life is like a play, with many entrances and exits by different actors, and the scene we see now is only part of the latest act. As in a television serial, the present episode is understandable only in the context of the previous ones, and what happens in this one will affect what happens in the next. We have to take a short excursion into ecological theory to explain why this is so.

Extinction is a completely natural and very common part of the processes of nature, not only at the evolutionary level (p.10), but also in living populations. It happens whenever a local population of animals does not breed fast enough to replace the individuals that die. If there is another population of the same species nearby, it will probably export individuals which sooner or later may re-establish that species in the place where it was temporarily extinct. This happens most quickly in the kinds of animals which get about easily and have wandering habits, like some kinds of birds and insects. In unpredictable patchy habitats on large land masses, local extinction and recolonization happen to some species just about every year. But if the locally extinct population is a long way from its relatives, it may be many years before a new one is established. For species which live on islands, the chances of recolonization across water from an adjacent mainland are slimmer than for continental species which can be recolonized from, say, the next field or the next valley. The smaller the island, the smaller the population and the more likely it is that some chance catastrophe will entirely wipe it out: the further the island from any source of new colonists of the same species, the less chance the population has of being re-established. The processes of population dynamics are clearer on islands than in more complex mainland communities, and in recent years the study of islands has grown into an important new sub-branch of ecology, known as island biogeography.[16]

Islands are different from mainland areas of the same size, in several ways. The most important difference is that islands nearly always have fewer species than the same area of mainland, and on oceanic islands (those that have never been linked to a mainland) the list of species present is quite unlike that of the nearest source of colonists. The faunas of oceanic islands are not a random sample, a representative selection of those living on the nearest mainland; the species most likely to be present are the ones with good powers of dispersal, for example birds – as opposed to, say, frogs. The lists of bird visitors to well-observed islands such as Britain or New Zealand are very long, and they include many unlikely exotic stragglers, as well as the regular migrants. Some of them are recorded only as individuals at long intervals, but others eventually establish breeding populations and settle down to join the native fauna. Their chances of doing this depend on, among other things, how many of the same species arrive together, how many other species are already present, whether the island offers the right habitat, and whether all the suitable living-spaces in it are occupied. During periods when the number of extinctions among the resident species is high, for whatever reason, there is room for more colonists. The number of potential colonists arriving per year depends partly on how far the islands are from the nearest mainland: generally, the more remote the island, the fewer colonists. New Zealand counts as a large, remote island, which has relatively few species compared with, say, Tasmania, an island only a quarter the size but much closer to the Australian mainland.[17]

There are several possible explanations for these observations. The one that has gained widest recognition is the equilibrium theory of island biogeography, proposed in 1967 by two American ecologists, R. A. McArthur and E. O. Wilson.[18] According to their theory, the number of species living on an island is a dynamic balance between the continual extinctions and immigrations that occur on an undisturbed island over a long period of time. The greatest number of species will be found on a large island near the mainland; the smallest number on a small, remote island. Moreover, the number of resident species is not constant; immigrations and extinctions go on all the time, not necessarily at the same rate. There is very likely to be a difference between the number of breeding birds found in censuses taken several years apart, which is why the bird checklists of island countries have to be constantly updated. The difference in the lists of species found in two censuses is expressed as a percentage, and is called the faunal turnover. The larger and the more remote the island, the slower the turnover, so long as it remains undisturbed.

Birds colonizing an island nearly always find a different environment from the one they left; milder in climate, perhaps, and with fewer of their old enemies and competitors. Given enough time and isolation, and a large enough population to persist over a very long time, the descendants of the colonists will adapt to the conditions they find in their new home, becoming steadily less like their parent stock, or in fact like any other stock in the world. They become *endemic*, or unique, to that island or group of islands.

The New Zealand group of islands has been isolated for a very long time, and includes two that are relatively large. Hence we may expect to find that many of those birds which have managed to reach New Zealand will have established populations large and persistent enough to have become endemic, and that is what we do find. The latest estimate reckons that 46 (60 per cent) of the 77 native land and freshwater breeding birds of the region are endemic.[19] (The marine birds are generally more wide-ranging, and only 30 per cent of 80 marine species of the region are endemic.)[17] The proportion of endemics among the land bird species before man arrived was certainly much higher than that. The peculiar snag about a bird fauna so dominated by endemic species is its terrible vulnerability to sudden environmental changes. Endemic species, by definition, cannot be replaced from outside, and are especially susceptible to extinction on their own home ground.

The marvellous abundance and variety of New Zealand's bird life 1200 years ago had developed over aeons of isolation – not always in peace and quiet, since the Ice Ages must have been a very difficult time for many species, and fatal for some – but at least always in the very minimum danger of a sudden encounter with a hungry hunter. The oldest residents had never met the most lethal modern predators (snakes, rats, carnivores and man); those that came later may once have been used to having to defend themselves against predators, but the longer they stayed, the more they forgot the experience, and the more completely they discarded all precautions against it. In undisturbed, isolated New Zealand, that was good strategy: individuals who did not waste energy on defence against non-existent ground predators had more to spare for other, more important things – like making sure that their own young were better represented in the next generation than those of their neighbours. It was a short-sighted policy, perhaps, but we can scarcely expect birds to plan for the future when we do not ourselves. It worked beautifully until the time came when the island fortress was invaded, the secure isolation of millions of years was broken, and the end of that primeval forest world began. Then the most ancient endemic birds found themselves ill-equipped for a new struggle: their kind of slow population turnover and naive behaviour, which had been successful before, was now a passport to oblivion.

Why all birds are not the same

Every kind of bird that has survived more than a few generations must be well adapted to the environment it lives in. That is, enough of the individual birds living this year must be able to find food and avoid death for long enough to produce young that breed next year. But the world is a very diverse place, and contains very many different kinds of habitats, most of them occupied by some birds. The problems of finding food and of producing surviving young are different in different kinds of habitats.[20] At one extreme are the temporary, patchy places in the process of developing into something else, like a pond or a landslide being overgrown with weeds. At the other extreme are the relatively unchanging places like mature forests and oceans. We may expect the kinds of birds that live in such extremely different places to be different too, not only in appearance but in population dynamics as well.

Birds that live in changeable habitats cannot be sure where to go to find food; they may be lucky and find an unclaimed feast, but they have a high chance of being unlucky and finding nothing. So they tend to be small, grow quickly, and mature early; during feast times they can lay large clutches and produce many young which disperse rapidly, but they usually live only short lives. Birds that live in more stable habitats, where food supplies are more reliable, are likely to live longer. They are able to specialize in diet, to grow larger and postpone breeding until they are older; they can then lay small clutches and produce only few young each year, and invest a lot of time and care in each one. Mammals and other animals show the same range of kinds within a group. Theoretical ecologists label the small, shortlived kind as 'opportunists'; they invest their energy in short-term reproductive output rather than in long-term survival. The other kind are called 'equilibrium' species; their investment priorities are the other way round. Of course, there are many kinds of birds that are intermediate between these two extremes; in fact, if we knew enough about them we could arrange the members of every family of birds in order along a spectrum from one extreme to the other.[g] Natural selection in undisturbed New Zealand favoured the equilibrium strategy, which is best developed in the birds that have been here longest. The species which have colonized the islands since their isolation have been, almost by definition, opportunists, but the longer they have lived here, the further their descendants are likely to have slowly drifted to the other end of the spectrum.

The ancestors of the takahe probably arrived in New Zealand well before the Pleistocene, so they have had at least several million years to adapt to conditions here. They demonstrate the characteristics most favoured by natural selection in pristine times: a large, heavy body, loss of flight, long life expectancy and small families. Cynthia Cass.

Among the living New Zealand birds, the takahe (p.142) is a prime example of an equilibrium species. Adult female takahe weigh two to three kilograms; they make only one nesting attempt each year, with one to three eggs in a clutch (most of their relatives lay much larger clutches than this); individual territory-holders in the wild can breed every year for four to ten years and, in captivity, takahe can live up to 20 years. An example of an opportunist would be the South Island fantail at Kaikoura. Adult female fantails weigh about 6-9 grams; they commonly lay three, four or even five clutches a year, with an average of three to five eggs per clutch; but individual fantails live very short lives. Fewer than one in a hundred survive long enough to breed in more than one season.[22] However, the distinction between opportunists and equilibrium strategies is not hard and fast, and some of the more adaptable species are able to vary their strategy according to their environment. For example, the South Island robins at Kaikoura have the high productivity and short life spans of opportunists, yet birds of the same species living in greater security on Outer Chetwode Island have lower productivity and longer life spans almost approaching those of the takahe.

These differences in overall life-history strategy will be important clues when we come to consider the profound environmental changes which were to take place in New Zealand. The changes had markedly different effects on different kinds of birds. The older endemic species, which had adopted, to various degrees, the equilibrium strategy, were by far the most vulnerable to change. Because they were large, each individual needed a large share of food and space, so in the restricted area of an island the total population had to be relatively small. Compared with smaller birds, they were naturally rare. Because they bred relatively slowly, they took a long time to recover from any interference which killed off too many adults – although conversely, they could easily cope with temporary disturbances which killed only the young, as the adults were long-lived and could breed many years in succession. Although birds which normally have a high productivity may breed more slowly in certain conditions, birds which normally breed slowly cannot greatly increase their output under *any* conditions: and these birds are usually the ones that have a very prolonged period of juvenile dependency, when the young are especially vulnerable to predators. Most important, the equilibrium species were unable to adapt to a rapidly changing environment. Each new generation of offspring contains individuals having different combinations of genes from those of their parents. The slower these new combinations are produced, the longer it takes for one to appear which is better adapted to its environment than its parents. Slow-breeding, long-lived birds were less likely to be able to produce new and different off-spring fast enough to keep pace with environmental changes, and so were more likely to become extinct. (Here, of course, we are not talking about local extinctions, which are only temporary and are more characteristic of opportunists.)

The vulnerability of large or specialized species to extinction is nothing to do with man's interference; it is a constant theme of evolution. Many of the fossil birds that we know about, from all over the world including New Zealand, were larger than their living equivalents; the moa were among the largest birds ever known. The same is true in other kinds of animals too. Like all generalizations, this one has very many exceptions, but it does reflect the generally greater risk of extinction run by the larger, equilibrium-type species. We should not be surprised at this. Only man is concerned for the survival of whole species; evolution certainly is not. Natural selection works mainly at the level of the individual, favouring those which are better adapted to conditions now and eliminating the rest. It prospers adaptations which profit the individual, whether or not they will profit the species in the long run. When they do not, extinction follows. The history of life on earth shows that this happens much more often than not.

The older endemic birds of New Zealand were unprepared for change in another, perhaps even more important way. Animals and birds subject to constant predation are wary, nervous, always alert even when at rest. Appropriate behaviour is part of the inherited equipment of a species in the same way that physical characteristics such as eyes and legs are, and for the same reason. The individuals who keep the sharpest look-out, distinguish approaching danger from the other less threatening things going on all around and react most quickly to it, are the ones most likely to survive to and through the breeding season. Other individuals, no matter how well adapted in other ways, will not have the chance to perpetuate their own characteristics if they do not get out of the way in time. The processes of evolution by natural selection will, over the generations, equip the species most easily hunted by predators with compensating watchfulness. But the process takes *time*, and if a hunter comes upon a population of animals unused to being hunted, it has a temporary but lethal advantage. If the prey are also endemic equilibrium species, such as are found on undisturbed, isolated, predator-free islands, then the predator will easily be able to remove them far more quickly than they can be replaced, and the resulting slaughter will be brief and final. In New Zealand, the older resident species had evolved on, or grown accustomed to living on, a forest floor free of such hazards; they had become practically defenceless – nature's equivalent of the babes in the wood, certain to be wiped out by the first passing wolf.

We know nothing about the behaviour of the various animals and birds, now long gone, which were living in New Zealand at the time the very first human colonists arrived. The early Polynesians could not of course leave any written description of them; but fortunately, there was one group of uninhabited Pacific islands which was still more or less untouched when an astute European naturalist visited it in 1835, and kept a detailed record of what he saw. The islands were the Galapagos, and the visitor was Charles Darwin.

Alan Moorehead tells us that Darwin found the Galapagos birds

. . . incredibly tame. Having never learned to fear man they regarded Darwin simply as another large harmless animal, and they sat unmoved in the bushes whenever he passed by. He brushed a hawk off a bough with the end of his gun. A mocking-bird came down to drink from a pitcher of water he was holding in his hand, and at the pools in the rocks he knocked off with a stick, or even with his hat, as many doves and finches as he wanted. On Charles Island Darwin saw a boy sitting by a well with a switch in his hand, with which he killed the doves and finches as they came in to drink; the boy told him that he was in the habit of getting his dinner this simple way. The birds never seemed to realise their danger. 'We may infer,' wrote Darwin, 'what havoc the introduction of any new beast of prey must cause in a country, before the instincts of the indigenous inhabitants have become adapted to the stranger's craft or power'.[21]

Only a few months later Darwin visited New Zealand, but the ship was homeward bound after more than four years away, and did not stay long. Darwin was violently homesick by this time, and probably not in the mood to realize that the early colonists he met were about to stage one of the world's biggest-ever unplanned predator-introduction experiments. With the hindsight regrettably afforded to us, we can see that there was a great deal of truth in what Darwin said, although we know now that it was the population biology as well as the instincts of the indigenous inhabitants that needed adaptation. Because the ancient New Zealand species were so handicapped in both respects, the events that follow unfold as a tragedy of predestined certainty.

2

THE POLYNESIAN PERIOD
8TH TO 18TH CENTURIES AD

The moa-hunters

About 1200 years ago[a] the first human adventurers discovered New Zealand. The tribal traditions of their descendants, the modern Maoris, tell of a Great Fleet of sailing canoes, each carrying a particular tribe, migrating from an ancient homeland called Hawaiki in 1350 AD. It was said to be a planned exodus, with the canoes stocked with everything needed to start a new life in a new land – seeds or potted plants of tropical crops including yam, taro and pumpkin (all originally from Asia), and sweet potatoes or kumara (from South America); dogs and rats for companionship and food; gardening tools, fishing and cooking gear. Unfortunately, archaeologists excavating early Polynesian settlements have found that at least some parts of this legend – including the date – must be wrong. For a start, they have found relics of a well-established hunting culture dating from at least 1100 AD.[3] Since it would take time to explore and adjust to a new environment, to learn new ways of finding food and keeping warm, it seems more likely that the first Polynesians stepped ashore around 750-800 AD. They probably came in small numbers, though whether by accident or design, or a bit of both, is unknown.

The tools, fishing gear, ornaments and burial customs of the earliest Polynesian settlers were similar to those used in eastern Polynesia in the period 500-1000 AD; and the languages of New Zealand and eastern Polynesia are closely linked. The most likely homelands of the original New Zealanders are the Marquesas or the Society Islands, some 5000 kilometres away across the open Pacific Ocean. Some authorities believe that a planned voyage of that length would have been far beyond the navigational abilities and known range of the Polynesians of that period;[4] they therefore conclude firmly that the first colonists to reach New Zealand must have come by accident. However, the Polynesians were quite capable of shorter voyages, and there is no doubt that they did, time and time again, set out into the unknown to find new land.

The inhabited islands of Polynesia are mostly quite small, able to support only so many people. A vanquished or oppressed group on one island may well have preferred the risk of an expedition to the conditions at home. J. M. Davidson[1] says that, to the Polynesians, the Pacific was full of islands, and the chances of making a safe landfall seemed high – at any rate, no one ever heard harrowing stories from those who had failed. So they may well have set out on purpose; but it seems unlikely that they had intended to go so far. The first people to see New Zealand probably arrived by a combination of intention and accident – an island-hopping journey that went wrong. Adrift after a storm, they could have been carried west by ocean currents, and then by the nor'westers of the Tasman to New Zealand. Whether there was a single canoe-load or several is unknown, but they all came from the same area,

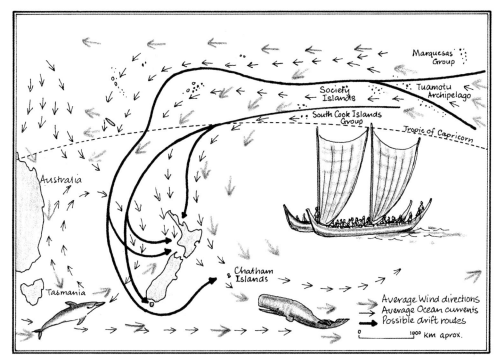

The South Equatorial ocean current and steady trade winds of the western Pacific would carry a voyaging canoe eastwards from the Marquesas until it reached somewhere near Fiji or the New Hebrides. If it then got caught in a tropical storm, it would be driven south to the Tasman sea, where the normal prevailing winds would blow it east to New Zealand. Cynthia Cass, after K. B. Cumberland.

and all within a short period. From that time until 1769, they had no significant contact with the outside world.[5]

Wherever they landed and however exhausted they were, the new arrivals would have had no trouble finding their first meal among the rocks, which must have been studded with large shellfish. When they ventured inland and saw their first moa, they may well have run away; but there were plenty of other large and meaty birds in the forest, many of them as flightless as the moa, and all of them tame and unsuspecting as well. In time they would have learned how to tackle moa too, the biggest prizes of all, and to trap them in swamps or bring them down with spears or harpoons. The hunt would have been dangerous and exciting, since the moa could probably defend themselves with powerful kicks from their massive clawed feet,[6] but there would have been plenty of incentive to persevere. Moa were relatively abundant[b] for such large birds (in the Eastern Polynesian language the name 'moa' is simply the word for the common fowl) and the meat of a single one of the larger kinds would sustain a sizeable group of hunters for several days. In addition to the birds, huge

The relative sizes of moa and man can be gauged from this photograph of Mr R. J. Jacobs working at Christchurch Museum on the reassembly of a complete skeleton of a giant moa (note the tiny skull, foreground) retrieved from a swamp at Pyramid Valley. A bird of that size, fighting for its life, would have been a formidable quarry for hunters armed only with stone weapons. Alexander Turnbull Library.

An excavation of an oven at a moa-hunter's camp in the Hawksburn Valley, Central Otago, in 1978-79. The dark earth on the right marks the burnt oven area – still containing moa ribs and other bones – in contrast with the lighter surrounding soil. The remains of at least 400 moa have been found at Hawksburn. Atholl Anderson.

concentrations of sealions and seals[c] hauled out along the coasts in winter, and the inshore seas teemed with fish. The new arrivals may well have thought that, although the climate of this strange new land was not what they would have chosen, the game and provisions were superb.

The most intensive hunting culture developed in the south: the largest and most recent (i.e. best-known) moa-hunter[d] sites are concentrated in Canterbury and Otago. The climate of those regions was probably warmer around 1200 AD than at present, but nevertheless, human lives in those days were hard and short – the average life span of adults was only about 30 years.[9] They often built their camps along the sea shore or at the mouths of large rivers, where they could hunt for seals, fish, shellfish and cast-up whales or dolphins, in between expeditions inland for birds. Some of the largest of these camp sites were occupied – or, at least, regularly revisited – for up to two hundred years. Their occupants threw onto their rubbish dumps volumes of evidence as to how they lived – just as we still do today. Archaeologists have found smoke-blackened oven stones, pits, hut sites and layers of greasy charcoal and shells; bones of swans, geese, rails, kiwis and eagles, and of 11 of the 12 different kinds of moa; and also bones of seals, dolphins, whales, rats, dogs, tuataras and fish. Local chieftains were buried with all their most valued possessions, including water bottles made from moa eggs, necklaces of moa bone reels and teeth of dolphins and dogs, knives, scrapers, awls and needles, fish hooks and ornaments of eastern Polynesian design, and a great array of different-sized stone tools for working wood. For some 500 years the moa and other large birds, and the visiting

A reconstruction of a moa-hunter burial, on display in the Canterbury Museum Hall of Pre-History. The body of an important person has been laid on its front, decorated with bone necklaces and supplied with the tools and equipment that might be required in the next life – including adzes and a moa-egg water bottle. Some archaeologists believe that the so-called moa-hunters did not rely on moa for the mainstay of their diet and could not have killed them in sufficient numbers to cause their extinction; but two hundred years of systematic collecting of moa eggs by sharp-eyed, wide-ranging and hungry Polynesian hunters, when combined with extensive habitat destruction, could have been the only form of predation necessary. Canterbury Museum.

Cultivation of tropical crops in a temperate environment was a difficult job, and one for which the Maori gardeners sought all possible help from spiritual, social and traditional sources. The hard work of turning over the soil with the ko *or digging stick was eased somewhat when it was done by teams of people moving together in time to appropriate chants and prayers. This picture was taken at the end of the nineteenth century at Ruatahuna in the Urewera, where traditional practices persisted into the age of photography.* National Museum.

seals, were the very basis of survival and civilization for a relatively dense and strongly carnivorous human population.

In the kinder climate of the North Island, the Polynesians managed to achieve – not without a struggle – a successful transition to a new, unique and substantial culture. They eventually overcame the formidable problems of getting their tropical food plants to grow in temperate conditions, and worked out how to store the produce over the winter. In due course they evolved the 'Classic' Maori culture, which was later 'caught alive' by the early European explorers.

Over the whole country, the way of life of each local group was, of course, strongly influenced by local conditions – there was never a typical 'Polynesian culture' common to all districts. Settlements depending on various combinations of gardening and hunting were soon widely distributed on both main islands, and from extensive explorations the people quickly got to know where all the important natural resources were. They travelled long distances to fetch the raw materials they needed: for example, a kind of black volcanic glass, known as obsidian, is ideal for making butchering tools such as knives and scrapers. Obsidian from Mayor Island, off the north coast of the North Island, has turned up in excavations of Polynesian settlements on both main islands. Greenstone from the south was known of in the north from early times, though it became important only later. By the thirteenth century the number of moa-hunters living in southern New Zealand was probably around 3000 people:[10] the total population of the country was perhaps 10,000.

In primeval New Zealand the forest was almost everywhere: of the total land surface, forest is estimated to have occupied 78 per cent, reaching to a treeline higher than today, and enclosing patches of natural grassland, swamps, pakihi *(a form of natural wetland, shown here). The only extensive areas of grassland were in parts of the dry heart of central Otago, where natural fires destroyed the forest about 2500 years ago, and on the 14 per cent of the land above the treeline.* Q. Christie, DSIR Soil Bureau.

The Polynesians knew how to make fire, but not how to control it. This sketch was made in 1919, but it gives a vivid impression of the fearful destruction of forest that must have gone on in the first few centuries of Polynesian occupation. There would be no escape from this for the flightless and weak-flying native birds. Alexander Turnbull Library.

Deforestation by the Polynesians

Untouched New Zealand was almost entirely covered in forest – at least, forest covered almost all land that was not above the treeline, or occupied by lake, sand or mire (Table 1). This enormous green blanket must have appeared to the Polynesians to be endless and totally indestructible. Maybe it would have been so, if they had been confined to attacking it with their stone axes. But of course they had another and far more terrible weapon. The native forests were particularly sensitive to fire, especially on the drier eastern side. Most of the canopy trees would be killed outright,[11] and few possessed any ability to resprout, or seeds that would remain viable for more than a short period. The understorey was dense, the ground litter thick, and drought was common almost anywhere except on the wet west coasts. The conflagration that could be produced after a dry spell by an uncontrolled fire through thick, tinder dry bush, fanned by the stiff winds common in New Zealand, would be quite sufficient to destroy vast areas of tall forest in a single holocaust.[c] Many of these early fires would have been accidental – or at least, larger than intended.

The first evidence of forest disturbance appears around 1000 AD in a few localities in both North and South Islands, but by and large the landscape at that time was still virtually untouched. By the late twelfth century, however, Polynesian settlements were widespread over the whole country, the population was rapidly increasing, and forest clearance was beginning on a grand scale. The lowland podocarp forests of the eastern South Island were destroyed between 600 and 800 years ago, and the beech forests of the eastern inland mountain basins disappeared between 500 and 700 years ago. Regeneration in these cooler, drier climates was slow or impossible, so the forest was replaced by tussock grassland. The vast open plains of Canterbury and

Near the Lochar Burn, at about 900 metres altitude on the Pisa Range south of Wanaka, a huge burnt log of Halls totara is evidence of the massive fires that raged across these hills in early Polynesian times. Similar logs from nearby have been dated by radio-carbon analysis to 1169 A.D. ± 49 years. The surrounding hills are now covered with fescue tussock grassland. A. F. Mark.

Otago look natural, but there are weathered stumps lying among the tussock, and layers of charcoal in the soil, which can be dated to the twelfth and thirteenth centuries AD.

Rapid deforestation was also underway in the North Island by 600-800 years ago, as we can tell from analyses of pollen grains preserved in bogs: the pollen of trees suddenly disappears from layers of peat and mud laid down after the end of the twelfth century, and instead pollen of bracken, scrub and grasses become more common. In the milder, damper climate of the fertile northern lowlands, much of Hawkes Bay and along the coasts, bracken grew quickly after fire, and within a few years could have been overtaken by second-growth forest, except that frequent new fires kept loading the odds in favour of the bracken. In the central North Island, forest that had not long achieved full recovery after the devastating Taupo Pumice eruption of about 130 AD was laid waste again by Polynesian fires. When the process of deforestation began, it was completed amazingly quickly – in some places the forest was totally destroyed within 20 years. Wherever the burned areas grew large enough to extend beyond the reach of seeds from untouched trees, or wherever the fires were frequent enough, the forest could not grow again. For several hundred years the destruction continued at a great rate; by the time it slowed down, in the late sixteenth century, millions of hectares of forest had gone (Table 1).

Later Polynesian culture

The basic forms of Polynesian economy – gardening, hunting and fishing, in various combinations – remained much the same throughout pre-history, except that after about the mid-fourteenth century the hunting of large birds gradually became more difficult. By the early fifteenth century, the most vulnerable game species (large birds

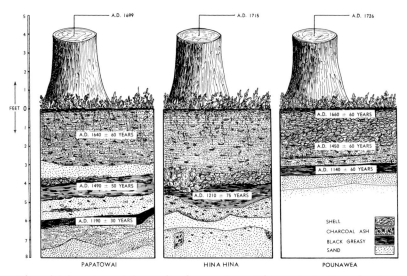

These sketches represent the results of excavations at three important moa-hunter camps in the South Island. At the lowest levels of all three there are numerous remains of moa embedded in black, greasy ash and charcoal, with radio-carbon dates starting in the twelfth century. By the middle or late fifteenth century the moa remains were being replaced by those of seafoods. The camps were abandoned in the seventeenth century, and all are overgrown by trees dating from soon after that. Leslie Lockerbie.

and seals) were very hard to find, and a period of cooler, wetter climate, causing a substantial advance of the Westland glaciers,[13] made life harder still for the hungry southern hunters.

As the climate and hunting conditions in the south deteriorated, the economic emphasis shifted more and more towards horticulture and, inevitably, settlements tended to concentrate in the warmer northern regions – just as they do today. By 1800, the total human population numbered around 100,000, more than 80 per cent of it in the north,[14] and the demand for food supplies was considerable. However, the forest no longer offered what was needed; the many kinds of vegetable foods found there were scattered, slowly replaced, or not particularly nourishing, and the birds that still remained were less vulnerable and more time-consuming to hunt.[f] By contrast, burnt-over forest in the north was usually replaced by bracken, whose enlarged overwintering storage root (*aruhe*) was an abundant and reliable source of carbohydrate.[17] The later Maoris respected the forest, though not, perhaps, in the way that some romantic reports would have us believe; but there was plenty of it, and it was not of much use to them. Besides containing by this stage relatively little food, it encroached on their gardens and villages, made travelling difficult, sheltered their enemies and, at night, contained a frightening assembly of imaginary creatures.[18] It was as much obstruction as refuge, as much enemy as friend, and constant burning-off of extensive areas to maintain the more useful bracken was simply the most obvious thing to do.[g]

The Polynesian rat and dog

The Polynesians themselves were not the only hunters to arrive during this period. They had companions, the first four-footed mammals ever to see New Zealand. The Polynesian rat or kiore originally came from South-East Asia; by 800 AD it had been distributed to almost all major Pacific islands by Polynesian voyagers, who regarded it as a delicacy. It came to New Zealand with them either by accident, hidden among food supplies in the bottoms of the canoes, or on purpose, to be itself a food supply for the voyagers. It is an agile and graceful climber, and it forages at night in the forest canopy as easily as a squirrel. It feeds mainly on fruit, seeds, and berries,[19] which is perhaps why its flesh was said to be sweet to the taste. However, like most rats, it will eat anything it happens to find, animal or vegetable, that does not offer too much resistance to being eaten.

During the Polynesian occupation of New Zealand the kiore was widespread on all the main islands, and was carried by human settlers to many of the offshore islands, including the Chathams. The hunters invented various kinds of ingenious snares and traps set across their runways to catch them, and observed strict ceremonies and rituals during the processes of capture and cooking.[17] No doubt they hoped that the correct repeating of appropriate charms would favour the hunt, or even bring on one of the occasional rat irruptions that meant plenty of food for all.

These bonanzas happened every three to five years, after heavy seedfalls of native trees. An abundance of seed allowed the kiore to increase suddenly (but temporarily) to very high numbers. The old-time hunters were well aware of this connection,[20] and that periodic swarms were always characteristic of kiore; in the South Island,

The Polynesian rat or kiore is a small (60-80 gram), clean, rather dainty little creature, with large ears and a long tail – rather like a cross between a ship rat and a mouse. Its fur is long and silky, brown on the back and white or grey on the belly. The Maori hunters were skilled at making elaborate traps or snares for kiore during the winter hunting season: one is shown in the background. Cynthia Cass.

these irruptions continued to the 1880s, that is, into the era of reliable recording. The irruption of 1884-85 in Nelson and Marlborough was described in graphic detail by J. Meeson:[21]

> The magnitude of the plague is the subject of ordinary conversation. . . . Living rats are sneaking in every corner, scuttling across every path; their dead bodies in various stages of decay, and in many cases more or less mutilated, strew the roads, fields and gardens, pollute the wells and streams in all directions. Whatever kills the animals does not succeed in materially diminishing their numbers. Fresh battalions take the place of those slaughtered. . . . A young farmer the other day killed with a stout stick two hundred of the little rodents in a couple of hours in his wheat field . . . I believe this is the true Maori rat . . . the Kiore Maori – the rat with whose presence we are now so largely favoured.

This was certainly a post-seedfall irruption. Meeson himself did not realize there was any correlation; but he mentioned that the beech trees had been seeding abundantly, and that the rats that visited the lowland districts had originally come from high up in the Inland Kaikoura Range, which was then still covered with beech forest.

The original Polynesian dog or kuri was similar to the common form of domestic dog in Polynesia, and was deliberately brought from there at some stage during the Polynesian settlement of New Zealand. In 1772 it was described as follows:

> The dogs are a sort of domesticated fox, quite black or white, very low on the legs, straight ears, thick tail, long body, full jaws, but more pointed than that of the fox, and uttering the same cry; they do not bark like our dogs. These animals are only fed on fish, and it appears that the savages only raise them for food.[22]

This is a detail from a picture by J. A. Gilfillan, 'Interior of a native village or "pa" in New Zealand', painted in about 1851 near Wanganui. The Maori subjects are obviously highly romanticized, but the kuri or Maori dogs look very much like those mentioned by the early explorers. Their descriptions agree on many details, including the medium size, foxy appearance, erect ears, short legs, and bushy tail. These dogs have all those features, and in addition look much more life-like than the few stuffed specimens that exist. Alexander Turnbull Library.

Another explorer reported in 1773 that the flesh of the kuri was indistinguishable from mutton; he added that the dogs were so dull and stupid as to be equivalent to sheep in mental capacity as well.

The kuri was prized and patted, used in hunting expeditions for ducks and kiwis, and as a household pet. It was also ceremonially killed and eaten by important persons, or offered up as a religious sacrifice; on Stewart Island, heaps of kuri jawbones have been found on the site of an early Maori settlement, and the name kuri lives on (unlike the pure-bred animal) in the Maori language and in several place-names. Perhaps most important, since there were no other mammals of any size available except an occasional seal, the hide of the kuri was the only source of leather for making clothes. The skins were cut into narrow strips, which were arranged so as to vary the colours attractively, and then either sewn firmly together or fastened side by side to a woven backing of flax. Early European visitors to New Zealand commented that the Maori seemed very fond of their dogs, keeping them tied with strings round their middles. It seems unlikely that the kuri ever became truly independent, at least not in any numbers, although the short-legged, bushy-tailed, jackal-like animal seen by Cook's party at Pickersgill Harbour might have been a wild kuri.[23]

Impact on the native fauna

For the native animals and birds, the consequences of the settlement of New Zealand by Polynesians, accompanied by rats, dogs and fire, were absolutely catastrophic. During the 1000 years that the Polynesians had the islands to themselves, at least 32 species of large birds, mainly rails and waterfowl, became totally extinct. They include all 12 species of moa, the pelican, the swan, two geese, three ducks, the coot, the harrier hawk, both eagles, the giant rail, four other rails, the owlet-nightjar and the crow.[24] In addition, three that still survive but which were originally common in both the North and South Islands (the takahe, the kakapo and the little spotted kiwi) became drastically reduced in distribution well before the arrival of the Europeans (Table 2).

Small birds and animals were not exempt from the turmoil although, because they needed less room, some could survive on undisturbed offshore islands. The tuatara, the New Zealand snipe, the Auckland Island merganser, and Stephen's Island wren were already gone from the mainland when the Europeans came. The mainland ranges of several species of large land snails became reduced; for example, the shells of extinct populations of *Paryphanta* and other, more strictly arboreal land-snails have been found in caves and swamps far from their nearest living relatives.

It is true that a few native birds – mostly those that had colonized within the previous thousand or so years – actually benefited: harriers and kingfishers probably gorged on small creatures fleeing the frequent bush and grass fires, or found easy pickings among the ashes; pukeko and native pipits rejoiced in the extension of open space. But these were the exceptions. For the rest, it was all doom and gloom.

Were these profound changes in the native fauna all caused by the Polynesians? After all, their population was small compared with the size of the country, and unevenly spread; and surely their stone-age technology could make only limited impact on the huge unoccupied hinterlands of their settlements. Such thoughts have

The large and impressive moa were not the only native animals to have been affected by the immigrant killers of the Polynesian period. By 1800 the New Zealand snipe and Stephens Island wren were already confined to offshore islands, probably driven off the mainland by kiore and bush fires. Subfossil bones prove that they once did inhabit both main islands. The giant native snails, Paryphanta, *no longer live in large areas of their former range, where their empty shells can still be found. Drawings not to scale.* Cynthia Cass.

prompted some writers to suggest that the extinctions of the Polynesian period have a natural explanation: the moa and other large birds were dying out naturally, long before man came. Two possible reasons for this view have been put forward.

The first is that there have been considerable changes in climate and vegetation since the most recent retreat of the ice – more subtle than those caused by a full glaciation, but still substantial enough to affect populations of birds. For the first three or four thousand years, the open grasslands of glacial times were overrun by the returning forests, and the warmest period was reached about 4000 years ago, since when it has been generally cooler. Perhaps these changes were sufficient to affect the populations or habitat of the large native birds so that, for example, the larger moa were less abundant by the time man arrived than formerly.[25] During the last thousand years, there have been several fluctuations in mean temperature, usually amounting to a rise or fall of less than one degree Celsius, but sufficient to cause substantial advances or retreats in the glaciers of the Southern Alps, and measureable changes in the steady temperature of caves.[13]

Without denying that these climatic changes could have had a considerable effect, these are two reasons why they are not, by themselves, sufficient explanation of the disappearance of the moa and other large birds associated with them. In the first place, the post-glacial adjustment in climate was nowhere near as severe a test of these birds' adaptability as the glacial periods which they had already survived. There is ample evidence that most of the species of extinct moa, rails, waterfowl and other large birds lived through all the glacial and post-glacial upheavals of the last two million years, and then succumbed during the latest one thousand, after almost all the natural disruptions were over. Short-term climatic changes in the same period, for example the brief cooler spells of 1050-1100 AD and 1250-1400 AD, may have made life difficult for the moa, and perhaps rendered them especially vulnerable to disturbance, but their final disappearance was too abrupt to be attributable to climate alone. Some other agent must have finished them off.

In the second place, climatic change severe enough to cause mass extinctions must have been worldwide. Indeed, New Zealand is not the only country in the world in which very large animals have survived two million years of profound environmental changes caused by the oscillating glacial/interglacial periods, only to succumb within the last few thousand. In Africa, America, Europe and Asia, the story is the same. However, these extinctions did not occur at about the same time, as would be expected if they were caused by some global change in climate. In Africa, the extinction of large animals was most marked about 40-50,000 years ago; in the Americas and Eurasia, about 10-13,000 years ago; and in New Zealand and Madagascar less than 1000 years ago. These dates all coincide with the arrival of human hunters in those areas, with weapons capable of tackling the larger species. As a result, 40 per cent of the genera of African large mammals, and 70 per cent of those of North America, disappeared.[26] For the New Zealand moa, the avian equivalents of the large mammals, the loss was 100 per cent.

The second suggestion is that species can become senescent, as individuals do, and lose their ability to cope with life. Some of the older ornithologists used to make statements like 'Some at least [of the extinct species] seem to have reached a kind of old age before they disappeared'[27] or 'Doubtless a large proportion of the species that have suffered most severely are forms that had lost much of their original vigour and were gradually dying out'.[28] This is the theory of orthogenesis, which states that evolution proceedes in straight lines that natural selection cannot regulate; certain trends, once started, cannot be stopped even if they lead to extinction – for example, the massive antlers of the extinct Irish elk.[29] Hence, one explanation for the demise of the moa presupposes that 'evolution had carried on their tendency to increase in size until their bulk must have become a burden to them, not only because of an increased demand for food but also because of their lessened powers of speed. So the moa came to be a misfit as a result of an unvarying evolutionary history, and was quite ready to succumb to a change of environment'.[27] Such a view of extinction implies that the extinct species had 'failed' in some way, but this is quite wrong.[30]

As we have seen, many of the ancient endemic birds of New Zealand, including moa, did tend to become very large, because in the undisturbed environment of those times, large size was favoured by natural selection. But the reason that the moa could not cope with the changed conditions after the arrival of man was not primarily that they were large, hungry or slow-moving, although they were all those things, but because large size and slow breeding are both characteristic of equilibrium species, which are generally unable to adapt to rapid changes in their environment. They did not lack 'species vigour' in the primeval habitat in which they had evolved – they had been successful there for millions of years. It was just that the changes brought by man exceeded their capacity to respond. It is true that the final stages of extinction may involve genetical changes which become irreversible in small populations, so that beyond a certain diminution of numbers, extinction becomes more or less inevitable.[31] But that is not an explanation of extinction in populations which had previously been large and vigorous: for them, we still have to explain what caused the initial decline.

The two most sweeping changes that the Polynesians introduced (the destruction of large areas of forest by fire, and the direct predation on birds) occurred more or less simultaneously, and their effects are strongly inter-related. Ideally, we would like to be able to distinguish them, at least to the extent of roughly estimating how many extinctions would have been inevitable, purely from the restriction of the area of forest alone. There is a correlation between the numbers of resident species and the area of their habitat, which is now very well established, holding true for a great variety of real and habitat islands, even to the number of mites on a mouse.[32] The implication is that, if the area of a habitat is reduced, some reduction in the number of species resident there is inevitable. Perhaps the theory of island biogeography (p.30) can help us to estimate the number of bird species that would have become extinct if deforestation were the only important explanation of their loss.

The total number of bird species present in New Zealand when the first Polynesians arrived is of course not accurately known. P. R. Millener's analysis of the history of the avifauna of the North Island showed that at least 73 species of land and freshwater birds have lived there at some time since the Ice Ages, and have left fossil or subfossil remains proving that nearly all of them survived into the early human period. So when the Polynesians arrived in the North Island, they found about 10.9 million hectares of forest and about 73 species of birds (Table 3). By 1800-1840, the forest was down to 8.4 million hectares (a reduction of 23 per cent) and the birds to 52 species (a reduction of 29 per cent). But according to the theory only ten per cent of bird species should be lost after a 50 per cent reduction in forest area, so we should expect a loss of 23 per cent of forest to reduce the number of birds by less than five per cent. Although this theory has many flaws (p.109), it does suggest, on this rather gross level, that the pre-European period saw many more bird species disappear than would be expected from the extent of deforestation alone.

One of the many troubles with this sort of historic calculation is knowing whether or not to include the waterfowl and grassland species which did not all depend directly on forest for their livelihood. It is true that in primeval New Zealand the forest was almost everywhere (Table 1), so very few birds could have lived far from forest of one sort or another; even the moa, usually thought of as being grassland birds, were probably quite at home in the forest (p.25). Few species could have been immune to the effects of the massive burning that the Polynesians practised. Deforestation affects not only forest birds; for example, pollution and flooding of formerly clean-running streams by accelerated erosion, caused by burning higher up in the catchment area, can seriously affect populations of inland waterbirds which seldom enter the forests themselves.[33] So, despite reservations about the theory, the conclusion suggested by the figures seems reasonable: the mass disappearance of native birds in Polynesian times was on a far greater scale than can be accounted for by the destruction of their habitat alone. Only one other major disturbance can account for it – predation.

Most of the 32 species of birds that became totally extinct throughout New Zealand during Polynesian times were large, which means that they would certainly have been very attractive to Polynesian hunters. In fact, of 27 mainland species known to have disappeared before the Europeans came, 25 have turned up in excavations of the moa-hunters' campsites and middens.[34] The earliest moa-hunter tribes relied very heavily on the meat of game birds and seals, especially in Otago and Canterbury. The most obvious and spectacular targets of those times were the dozen or so species of moa, of which all but one are known to have been killed and eaten by the moa-hunters. According to A. Anderson,

> the predator-prey equation of moa and man stacked the odds very heavily in favour of the hunter. The prey was flightless, comparatively slow moving and probably conservative in its territoriality and choice of breeding sites ... it was, moreover, almost totally naive to predation – perhaps only the eagle was a threat to moa young – it had no refuges inaccessible to man and, beyond kicking, it had no natural defences. By analogy with emu and cassowaries it may also have had a long incubation period, during which the male was virtually immobilised on the eggs ... To these behavioural disadvantages can be added the crucial fact that moa were by far the largest flesh "packages" found in the terrestrial environment. On the side of the predator lay the existence of a perfectly capable pre-adapted hunting technology ... and a lack of familiarity with sustained-yield culling of large terrestrial game.[35]

The archaeological evidence shows that, in some inland and southern areas, very large numbers of moa were killed in the first few centuries of moa-hunter occupation. One site on the Tahakopa River in southern Otago, in the extreme south of the South Island, yielded the remains of 678 moa eaten there between about 1200

What may well have been the last moa-meat dinner ever was eaten in a rock shelter under a limestone bluff across Takahe Valley from this one, in the Murchison Mountains, Fiordland, about 1700 A.D.[38] In 1949 the scorched remains were found by K. Miers, a Wildlife Service officer working on the newly rediscovered takahe. The dry dust at the back of the shelter had preserved bones and feathers of moa, takahe, kiwi, weka, and kakapo, as well as a still usable sandal of plaited flax, flax nooses, a lure made from weka tail feathers (see p.76) and a pointed wooden fire-stick.[3] Author.

and 1475 AD.[36] Moa bones dating from the mid-twelfth century AD were so numerous at the mouth of the Shag River, in Otago, that they were dug up, crushed and sold as fertilizers.[37] Remains of Moa eggs have been preserved less often than bones, but they too were taken in very large numbers during the breeding season. Small wonder that, after about 1350 AD, moa-hunting declined; very few moa were left in Canterbury and Otago by 1450, and moa-hunting as a way of life seems to have finished in most of the south by the 16th Century.[38] Most of the smaller game birds seem to have disappeared even earlier. Archaeologists know of no remains of the extinct swan, goose, duck, goshawk, coot and crow dated after 1200, or of the giant rail after 1350. The final blow for these ancient innocents did not necessarily come from a stone axe: the burning of their habitat must have helped to weaken their populations or expose them to the hunter's eye, and the persistent theft of their eggs would have drastically reduced their birth rate, which would in time have finished them off even if the adult moa were immune to hunting. Either way, it seems entirely reasonable to attribute the final extinction of all these tame, meaty birds to the actions, direct or indirect, of human hunters.

Some authorities have objected to the idea that hunting could have been to blame, maintaining that the moa hunters had a primitive system of conservation to husband their prey.[39] Others point out that, although the human population was unevenly distributed in the two main islands, the rate of extinction of birds in pre-European times was about the same in both.[31] Both these objections may be valid, and yet the conclusion that hunting was the major factor still stands – not because of the attributes or behaviour of the hunters, but because of those of the prey. The large endemic birds were utterly incapable of adapting, either as individuals or as populations, to being hunted by any predator, least of all by organized groups of hungry, agile and intelligent men. If the moa-hunters had a conservation ethic, it was not very effective: more likely, the European writers who enthused about the reverence for the forest felt by the Polynesians were rather romanticizing what they saw, either because they looked for proof of philosophical ideas about the 'noble savage',[40] or because the early Maoris (who loved a joke, especially at the expense of gullible and unsuspecting visitors)[41] were simply leading them on. The plain fact is that the Polynesians did not live in harmony with their universe, either in New Zealand or on other Pacific Islands (Chapter 6); their impact on the landscape and its biota was severe, whether by intention or accident, and the 'reverence' they felt was not equivalent to, and should not be confused with, modern ideas about conserving natural environments for their own sakes alone.

The effects on the bird life of the two other immigrant killers present during this time, the Polynesian dog and rat, are uncertain. The true Polynesian dogs were highly prized possessions, carefully kept and bred for food and for their skins, so they were probably mostly domestic. The kiore is a small, mainly frugivorous rat, an agile climber which can forage with equal ease in trees or on the ground. Modern observations show that the birds now affected by kiore are mostly seabirds, plus some tree-nesting land birds, but no ground-nesting landbirds,[42] so the kiore has gained a reputation for being 'harmless' to native birds. But this is largely because kiore are now mostly confined to offshore islands where seabirds are most common. It does not imply that kiore had no effect on the very abundant ground-dwelling birds of a thousand years ago, and especially on their eggs and young – not to mention various other animals that then lived or nested on the ground.

In modern times the wildlife of offshore islands with and without kiore are quite different; some small petrels and shearwaters, tuataras, lizards and some invertebrates and plants are much more abundant where there are no kiore.[43] On Motuara Island over three summers (1974-5 to 1975-6) kiore destroyed 26 per cent of 31 robin nests being observed, including some that were 12 metres up in the canopy.[44] It seems quite possible that the widespread pre-European kiore population could have had a substantial effect on the smaller mainland birds which laid small eggs in nests on the ground, and could even have caused extinctions of small forest birds, unknown to science, that left no subfossil remains.[h] The huge numbers of kiore reported in the summers after heavy seed-falls would certainly have been very damaging: in contemporary times, kiore are known to be capable of reaching very high numbers on their island refuges (p.151). Although they may be less effective predators than ship rats, their effect on breeding birds at such times could still be significant, and they had hundreds of years to exert it in. The evidence is certainly rather slim, but it is surely more than enough to suggest that the kiore may not have been quite so 'harmless' in the past as it may appear to be now.

Conclusion

The distance in time and everyday experience between the world of the moa-hunters and the world that we live in today is vast; the two could be on different planets. We may never fully understand their lives and the way they and their environments affected each other; we may never be able to tell whether it was their direct action that exterminated so many birds, or some external factor that would have caused the birds to disappear anyway. By and large, though, I think the 'overkill' hypothesis is more convincing. It seems clear that the combination of direct predation by Polynesian hunters on the adults and especially the eggs of the larger native species, and by kiore on the smaller species, plus the effects of extensive man-made fires, must have been completely devastating for the naive native fauna. Nothing in its evolutionary history had prepared it for such an attack, and it was this pathetic vulnerability, more than any particular ruthlessness on the part of the invaders, that made the impact so inescapable. The circumstances of the encounter guaranteed that the immigrant killers of the Polynesian period were terribly destructive, especially at first. This does not deny that climatic change may not also have played some role, especially for takahe (p.145) and some moa; it is simply that whatever environmental changes were due to climatic change were significantly overshadowed by those set off by man.[45]

Nevertheless, there were also huge areas of forest that remained virtually untouched, which continued for hundreds of years to support large populations of those native birds that could come to terms with the attentions of stone-age hunters by day and of kiore by night. They might have continued thus indefinitely, if they had been left alone: but the eighteenth-century world was shrinking rapidly. The beginning of the end for the entire island community – hunters and hunted alike – would, in the course of time, be determined by events in the great European seafaring nations on the other side of the world.

In March 1773, after four months in the paralysing cold of the Antarctic, Cook rested his crew in Dusky Bay, in Fiordland. He brought the ship so close inshore that a tree growing out horizontally across the water reached the gunwale and served as a gangway for the mariners – and for their animal companions. Every ship of the period was infested with rats, and for that reason also carried cats: both would have been delighted to take to the bush, especially as it was stocked with unwary birds. This painting by William Hodges illustrates the scene described in the extract from Cook's journal quoted on p.68. National Maritime Museum, Greenwich.

3

THE EARLY EUROPEAN PERIOD
1769–1884

The coming of the Europeans

The first Europeans to set eyes on New Zealand were the Dutch crewmen of Abel Tasman's ships, the *Heemskirck* and the *Zeehaen*, which anchored in Golden Bay in 1642.[1] Though the Dutchmen were friendly, they had no way to convey their intentions to the suspicious and hostile Maoris who paddled their canoes out to meet them. A Maori canoe rammed a jollyboat, which was passing between the two ships, and the warriors attacked its crew. Four men were killed, and Tasman beat a hasty retreat, pursued by more canoes launched from the beach. The visit was brief and ill-fated; none of the Europeans, and probably none of their animal companions, set foot on the land. When they had gone, New Zealand and its people were left undisturbed for another 127 years, until a new phase for both was heralded by the arrival of the British explorer Captain James Cook.

Cook was a humane and observant man, a dedicated seaman and a highly skilled navigator. He made three visits to New Zealand, in 1769-70, 1773-74 and 1777, accompanied by a number of scientists and artists including Joseph Banks, Daniel Solander, J. R. and George Forster, Sydney Parkinson, William Hodges and John Webber. Their descriptions of the land and its abundant natural resources soon attracted many more European visitors, some to grab what they could and get out, some to settle permanently. They found a Polynesian society with distinct regional variations, living in a country larger and more diverse than most of them appreciated at the time. The best-known picture of it was based on the Bay of Islands, and the Maoris living there were taken as typical. Europeans arrived early in this area, and visited in some numbers; their influence there was described – sometimes for political or idealistic reasons – in emotive terms (the 'fatal impact' theory), and generalized to the whole country. In fact, there were some areas where Maori society quickly adapted to European ways, and others where the pure Maori culture of the pre-European era remained almost unchanged for another 50 or 60 years.[2]

The main interest of the first wave of European visitors was the rapid exploitation of a few basic items reported by Cook and his party – especially seals and whales, timber and flax. Sealing began in Dusky Sound in 1792, when a gang left by the *Britannia* collected 4500 skins in ten months.[3] The sealing ships came from Sydney, Hobart, the United States and Britain, and concentrated on the south coast of the South Island, especially in the Foveaux Strait area between 1803 and 1810. Seals had been hunted by the Polynesians in moa-hunter days, but the motives and methods of the European sealers were different. There was an international demand for sealskins, seal oil and felt, and the cost of fitting out an expedition was relatively small compared with the size of the profit that was possible. Commercial greed and international rivalry ensured that the Europeans depleted the stocks in a fraction of

Sealing was dirty, brutal work but, for a short period in the early 1800s, fur-seals were apparently innumerable, concentrated in convenient herds, and harvested as easily as men 'kill hogs in a pen with mallets'.[4] This sketch was probably made on the Auckland Islands soon after 1810; by 1830 there were practically no seals left there. Alexander Turnbull Library.

the time taken by the moa-hunters.

The sealers worked with 'reckless efficiency' to put themselves out of business. The first sealing gang to land on the Antipodes Islands, 790 kilometres east of Stewart Island, collected in a few months a haul of 60,000 sealskins;[5] seals were almost exterminated on the Antipodes within 20 years. In 1810 the sealers discovered the subantarctic Campbell and Macquarie Islands, and transferred their attentions there, with the same effect. In 1830, an American sealing captain visited the Auckland Islands in search of seals, but he was disappointed. He wrote:

> Although the Auckland Isles once abounded with numerous herds of fur and hair-seal, the American and English seamen engaged in this business have made such clean work of it as scarcely to leave a breed; at all events there was not one fur-seal to be found on the 4th of January 1830. We therefore got under way . . . and steered for another cluster of islands, or rather, rocks, called 'The Snares', 180 miles north of the Auckland Group and about 60 south of New Zealand . . . we searched them in vain for fur-seal, with which they formerly abounded. The population was extinct, cut off, root and branch, by the sealers of van Diemen's Land [Tasmania], Sydney, etc.[6]

After the 1830s sealing was not profitable alone, and became a supplement to whaling, farming or trade; the abandoned sealing camps were left to be cleaned up by the rats.

Deep-sea whalers prospected New Zealand waters in 1791-2, and were well established by 1802. Unlike sealing, deep-sea whaling needed a lot of capital, and most of the ships were American or British. In 1839 there were, on one estimate, 80 American whaleships operating in New Zealand waters. But after 1840 they moved on, as the effects of excessive harvesting began to show, new whaling grounds opened up elsewhere, and the new British administration began to impose customs duties on foreign vessels.

Both before and after the ships left, whaling was also carried on from shore stations, mostly in the South Island and on the Kapiti coast. The season was from May to October, when the cow whales came inshore to give birth. The whalers would row out in their sturdy boats and, whenever opportunity offered, kill the calf so as to trap the cow, which would not desert her young. This 'unprofitable and cruel proceeding' inevitably destroyed its own livelihood, just as sealing had done before it.[7] But, as with sealing and deep-sea whaling, there were additional reasons for the decline of the shore-based whaling industry, including economic depression, better alternative investment opportunities, the use of cheaper vegetable oil in place of whale oil, British tariffs and the changing political conditions after the British annexation.

The masters of various visiting European sailing ships soon realized the value of the abundant natural resources of New Zealand. The leaves of the native flax (quite a different plant from the true flax from which linen is made), when dressed in the traditional way by patient Maori women working with mussel-shell scrapers, yields a strong fibre suitable for making into ropes and cordage. In the forest there were several species of trees that grew tall and straight, ideal for making masts and spars, and also many other species producing timber of different properties for other uses. Traders dealing in flax and timber profited from the eager co-operation of the

When a whale was sighted, boats' crews from several ships would compete to 'make fast' their harpoons in it first. Many deep-sea whaling ships worked the seas around New Zealand during the first 40 years of last century, and left their legacies of Norway rats and half-caste children all round our coasts. Clark's painting dates from 1819. Alexander Turnbull Library.

Men from the shore whaling station at Te Awaiti, in the Marlborough Sounds, pose in their four-oared harpoon boats. The whaling station behind was also a farm, with fences made from whale's bones and the bush on the hills behind burnt off for pasture. Norway rats must have bred up to great numbers in such places, and from there spread into the interior. Canterbury Museum.

The extensive stands of kauri in North Auckland were a timberman's paradise. Such a concentration of enormous trees, with trunks as tall and straight as cathedral pillars, and smooth-grained, knot-free, silky wood, offered visions of profit that no one could refuse. Systematic, large-scale kauri milling was a dominant industry between 1820 and 1870, and smaller-scale operations went on much longer (this photograph was taken in 1918). Logging and fires have removed over 99 per cent of the original three million acres of kauri forest (10,000 acres are left) along with all the rich native fauna it used to support. Alexander Turnbull Library.

Maoris, who at that time were willing to make almost any sacrifices to trade for muskets, gunpowder and iron tools. Aggressive Northland tribes armed with these devastating new weapons were causing havoc around the Bay of Islands well before 1830; during this time many settled offshore islands (e.g. Little Barrier and Hen Islands) were totally depopulated, and have remained so ever since.[8] Such raids terrified the other tribes, so the women laboured over the flax leaves, and the men were prepared to 'pull and haul and sweat like Horses'[9] to drag timber out of the bush.

The exploitation of essentially one-crop timber trees, such as the Northland kauri, was as ruthless as that of other resources – often leaving a 'melancholy scene of waste and destruction'.[10] On the other hand, Maori farming was greatly stimulated by the early introduction of the potato, an easy crop to grow compared with kumara, and profitable too, as there was always a ready sale for this staple food among the Europeans. The Maori also learned many other new agricultural skills from the missionaries, and were introduced to pigs, cattle and horses; by the late 1830s they were trading in pork, barley, oats, peas, maize and wheat, providing much of the food needed by the Europeans of the Auckland region, and even exporting to New South Wales.

In the 1860s came another wave of transient exploiters, with even less respect for the land than any that had come before. The gold rushes of 1861 and 1865 brought diggers in their tens of thousands to Otago and Westland. They tramped through regions of the thick, wet West Coast forests that had seldom been visited by man, and demanded huge quantities of food and building materials. Much of it was imported from Australia, but they also took birds and timber from the forest around them. In 1867 the human population on the West Coast reached a peak of 29,000, only two years after the first strike.[11] Their operations laid waste large areas of land – there was never any thought of rehabilitating the surface workings or the huge piles of dredge tailings. Inevitably, production declined within a few years, but even today, after 120 years and 28 million ounces of gold won,[12] some mining continues, long after the seals, whales and kauri have all been reduced to a few protected remnants of their former abundance.

It is hard for us to imagine the bewildering speed and depth of the changes that overtook the Maori population in the first hundred years of European settlement in New Zealand. In that time, the Maoris living in the areas most exposed to European influence were catapulted from the stone age into daily contact with the products of the Industrial Revolution. They had to absorb new ideas, adjust to a new language and system of government and to changes in dress and diet, and cope with the effects of alcohol, firearms, money, and unfamiliar diseases – especially influenza, tuberculosis, dysentery, measles, bronchitis, typhoid and whooping cough. In short, they had only a few decades to make cultural adjustments which the ancestors of their visitors had taken centuries to pass through. Small wonder that in parts of the country, such as the Bay of Islands, the process took a terrible toll in social dislocation and increased mortality from war and disease. The humanitarian efforts of the missionaries often seemed to make matters only worse. The dreadful irony was that, despite their best intentions, the axes they provided turned out to be both useful tools and murderous weapons; blankets worn in the style of traditional Maori clothing had bad effects on health; and the ending of cannibalism left a serious gap in diet.[13] Places more remote from disturbance, such as the interior of the North Island, had time to absorb the changes more slowly.

These differences between areas led to different conclusions about the extent and impact of the European colonization. There is evidence both of the transformation and of the continuing stability of Maori cultural life – both were true, but neither was universal. But there was a general decline in the total population of Maoris, which continued to the end of the nineteenth century. Early reports referring to the 'depopulation of the natives' caused concern in Britain; it was especially obvious in the areas most exposed to change – where Europeans were most likely to see and report it. The total estimated Maori population in 1843 was still much as it had been in Cook's day – around 100,000;[14] but by 1857 it was down to 56,000, briefly equalling the rapidly rising number of Europeans. In 1896 it reached a low of 42,000, by which time Europeans outnumbered Maoris by seventeen to one.[15] Since then, the Maoris have steadily increased in numbers – but, of course, so have Europeans; Maoris still constitute less than ten per cent of the total population. The Maori way of life has been, until recently, almost completely submerged under the alien culture imported *en bloc* from overseas.

Until the Europeans arrived, Maoris had no name for themselves as a people, only a great number of tribal names. Now they began to use the term *tangata maori* to distinguish themselves from Europeans, by adding the adjective *maori*, meaning 'usual, ordinary' to the word for 'man'. Europeans used *maori* as a noun, and the Maoris eventually accepted this name for themselves. It is still a matter of debate as to when the Polynesian immigrants of 1200 years ago became the Maoris of today, but the origin of the term, as applied to the people and their culture, is no earlier than the time of first European contact. *Pakeha* is the reciprocal term they applied to the white-skinned invaders.[16]

The flow of European and Australian traders and settlers was uncontrolled but moderate until February 1840, when the Treaty of Waitangi made New Zealand a colony of Britain. After that, organized settlement increased by leaps and bounds. The settlers struggled to reproduce in the Antipodes a European-style economy with immigrants, capital and technology imported largely from Britain. To begin with, the Maoris welcomed the settlers and the trade and material progress that came with them. For many years they continued to provide the settlers with food, and were willing or even eager to sell land surplus to their new farming requirements, especially as the more astute among them saw the opportunity to deal in land as a new way to pay off old scores against other tribes.[17] Maoris also eagerly bought or traded for liquor, tobacco and other dutiable items, and so contributed generously to the customs revenue of the new colony. But, in the North Island, where most Maoris lived and which had the most attractive climate for agriculture, the settlers' increasing demands for land, and the Maoris' own complex system of ownership and ethics, led to disputes, and then to conflict. By the time the leading Maoris realised that the few traders and settlers they had welcomed were being followed by hordes of land-hungry European colonists, it was too late to turn back. The result was, inevitably, open rebellion, which drew a predictable response from the colonial government.[a] During the 1860s, the process of settlement of the North Island slowed down, even to a halt in the most hotly disputed areas.

Meanwhile, inter-tribal warfare had so lowered the resistance of the South Island Maoris that land purchase deals with the settlers went through with ease, and at ridiculous prices.[19] There was tremendous potential for sheep farming on the open plains of Canterbury and Otago (and also to a lesser extent in parts of the eastern North Island, such as the lowlands of Hawkes Bay) which had been already stripped of totara forest and scrub by the moa-hunters centuries before. But even these open and available ranges – the last unclaimed and unused temperate grasslands anywhere in the world[20] – needed some preparation before the sheep could be established.

The tussock provided a dense, wind-rustled blanket over the plains and up the hills to an altitude of over a thousand metres. Beneath the tall, bunched grasses, in the shelter they gave from drying nor-west winds and intense sunshine, grew other grasses, herbs, sedges and small flowering plants. These shelters were also the home of many small native animals and birds, such as the land snails and the native quail, and abundant insects. Scattered over the plains were clumps of shrubs, such as manuka and cabbage trees, the spiny matagouri, and a savage spear grass known as the 'spaniard'. Matagouri and spaniard caused great pain and trouble to the earliest visitors to the tussock country, like J. T. Thomson in 1857:

> The 14th of February found us camped on the banks of the Oreti [Southland] . . . The country here bears fine grass, but is much overrun with a scrub called Tomataguru [matagouri] by the natives, or 'wild Irishman' by the colonists. It is full of prickles and is

The leaves of the spaniard grass are tipped with rigid, needle-sharp spines. The flower-stalks are protected with similar spikes set to point in all directions. The 40 or so species of Aciphylla *in New Zealand include some, much larger than this one, which fully justify species names such as* horrida *and* ferox. *The reason they needed to be so well protected is unclear, but the consequences were bad for the fauna of the grasslands, which the settlers burnt to get rid of the spaniard.* Author.

difficult to penetrate. . . . On the 15th of February we proceeded to the foot of the Dome Mountain. . . . A new plant appeared here . . . . colonially termed a 'Spaniard'. . . . It has stout blades with sharp points – no agreeable object to encounter. The country here is much overrun with these and 'wild Irishman'; so much so that it was a matter of some difficulty to drag our horses through them, for the poor animals in swerving from the talons of the 'wild Irishman', were apt to be received on the more deadly weapons of the 'Spaniards' . . . The constant forcing our way through high grass, fern and scrub has worn shoes and trousers into holes and rags. We tumble dozens of times in a day, one time over a tussock, another time into a hole; now against a 'Spaniard', and then into the rough arms of a 'wild Irishman'; till our legs are raw with jags and scratches, and our hands and arms are full of thorns. The hair is even worn off the legs of our horses, and their fetlocks are full of sores. . . . On turning back we set fire to the grasses so as to give facility to future travellers . . . now fully 30 miles is in a blaze.[21]

For years after the arrival of the first white settlers, tremendous conflagrations swept the tussock country. These huge fires, fed by the accumulated plant debris of centuries, must have been exceptionally intense, but the settlers did not know that it was the small, palatable plants that grew between the tussocks that were the most favoured by the sheep, and also the most susceptible to very hot or repeated firing. They noticed only that burning cleared away the thorns and stimulated new growth in the tussock, so they burned the same land at frequent intervals – as much as every four years.[22] Their hardy Merino sheep, imported from Australia, increased rapidly in numbers until the 1880s. The pastoral runs of the South Island high country, usually covering 4000 hectares or more, very quickly became the economic backbone of the country during the 1850s and 1860s. Soon the wool barons were wealthy enough to build the gracious cities of Christchurch and Dunedin, each with large stone public buildings, including museums and colleges, while the business boom stimulated by the gold rushes continued unaffected by war. All over the country the cultural influence of the Maoris declined as that of the Europeans strengthened.

Deforestation

Forest clearance in the pre-European era had slowed down after the sixteenth century, probably because the Maoris had already destroyed the most vulnerable forests (mainly in the South Island) and, according to McGlone, they had no better use for those that were left (p.44) – at least, not until the introduction of the potato and other temperate-climate crops. The extent of forested land in 1840 was reckoned to be about 14 million hectares, covering some 53 per cent of the country – already greatly reduced, from the estimated 21 million hectares (78 per cent cover) of primeval times, by the ancient Polynesians (Table 1). With the coming of the Europeans, and the introduction of new crops and livestock, a new phase of forest clearance began, concentrated in the North Island.

At first, kauri and other valuable timber was clear-felled, and agriculture – especially by Maoris – was rapidly expanded. More extensive clearing, cutting into more difficult and less immediately profitable forest, was delayed by the land wars of the early 1860s. When they were over, about 800,000 hectares of Maori land was permanently confiscated,[b] and the government brought in 3500 new immigrants from Scandinavia and established them in special settlements, to tackle the huge area

Most forests cleared by the Europeans were burnt after only partial logging. The first stage of preparing for a burn was known as 'under-brushing', and involved cutting the fern, vines, supplejack, bushes and small trees and leaving them to dry. In the right conditions this flammable material carried the fire across the ground, and achieved a burn hot enough to consume the denser material that could not be removed any other way. Fire was therefore the main ally of the settlers, just as it had been for the Polynesians. William Strutt sketched this working party in 1856. Alexander Turnbull Library.

of untouched forest which lay across the hinterlands of Wellington, Taranaki and Hawkes Bay.[24] They pioneered the felling of the Seventy-Mile Bush in the northern Wairarapa, and showed how it was possible to convert solid forest into dairy farms. From the mid-1870s to the mid-1880s, timber was the colony's biggest industry; by the later 1880s, as dairying became established in the north, the decline of gold, wheat and wool prices in the south began to swing the balance of natural advantages of the two islands in favour of the north, where it has been ever since. The advantage was not, however, won easily.

Attacked with the metal tools and aggressive spirit of the nineteenth century, the North Island forest fell back rapidly, but for the individual pioneer farmers and their families, it was still backbreaking and heartbreaking work. It left them in no frame of mind to consider the long-term results of their labours for the original inhabitants of their newly adopted country. It was not that the results of extensive and rapid deforestation were entirely unknown; George Perkin Marsh's famous and influential book *Man and Nature*[25] was first published in New York and London in 1864, and reprinted in 1865, 1867, 1869 and 1871 and with second American and English editions in 1874. It is inconceivable that *no* settlers or policy-makers in New Zealand had read Marsh's graphic descriptions of the consequences of forest clearance for soil, climate and resident fauna; yet the destruction of the North Island forest went ahead anyway. As in North America, 'Romanticism was not allowed to interfere with the pragmatic aspects of making the wilderness productive . . . the settlement process in North America . . . can be viewed as a series of environmental traumas or conflicts, in each of which the modern American has won the immediate decision through a technological knockout.'[26]

Life for the pioneers was, by our standards, pretty grim – and they in turn made it grim for the native animals and plants they encountered. The living kauri forest around this gumdigger's hut has been swept away, and the pioneer's cat and pigs hunt among the remains; but there would have been little point in talking to this family about wildlife conservation, even if anyone had thought of doing so at the time. Canterbury Museum.

In the South Island, the established sheepfarmers were, at least to start with, much wealthier than the struggling northern dairymen. Few of them, however, had any great regard for the native wildlife, or any conscience about their destructive management practices, until the consequences began to show in economic terms. By the late 1870s individual second-generation farmers were being faced with the results of three decades of rural exploitation: the depletion of native pastures by overstocking, fire and erosion, and the spread of noxious weeds. But over the whole country, the total number of sheep and cattle continued to rise, since the decline of the earliest-established pastures was more than offset by the continued breaking-in of new farms from forest, and the conversion of native to European pasture.

The early settlers of New Zealand found a land which, in climate and natural conditions, seemed to duplicate many of the best features of the homeland from which they came, except that the native birds retreated with the forest in which they lived, and there was absolutely no game except the fierce and hairy wild pigs. G. M. Thomson, one of New Zealand's most prominent early scientists, wrote:

> Here, in a land of plenty, with few wild animals, few flowers apparently. . . . with streams almost destitute of fish, with shy songbirds and few game birds, and certainly no [native] quadrupeds but lizards, it seemed to them that it only wanted the best of the plants and animals associated with [their memories of their homelands] to make it a terrestrial paradise. So with zeal unfettered by scientific knowledge, they proceeded to endeavour to reproduce – as far as possible – the best-remembered and most cherished features of the country from which they came. . . . They recked not of new conditions, they knew nothing of the possibilities of development possessed by species of plants and animals which, in the

severe struggle for existence of their northern home had reached a more or less stable position. . . . No biological considerations ever disturbed their dreams, nor indeed did they ever enter into their calculations. . . .[27]

So, over a period of about 60 years from the middle 1850s, the Acclimatization Societies rapidly established themselves, and began to introduce many different kinds of animals. The intensity of the enthusiasm for acclimatization, especially of game mammals, was partly cultural: 'The landed class wanted the familiar sporting animals, and the far more numerous members of the underprivileged classes, especially those who had lived for a generation in the relative freedom of pioneer life, were even more avid to enjoy the sport and food available to their fathers only at a poacher's risks. Indeed, many of the transportees had been sent to New South Wales as the price of a dinner of roast venison or rabbit stew. A wide variety of game, and freedom to take it at will, regardless of social and economic position – these symbolized for them (as could little else) a new freedom from galling class discrimination.'[28]

Besides those species brought for food or sport (probably the majority), there were others which were simply reminders of home. A surprising proportion of attempted introductions failed. Only 36 species of exotic birds are now established, of 130 liberated or escaped from captivity; and 33 species of mammals, of 51 liberated and three brought in by accident on ships – different authorities give slightly different figures.[29] Astonishing numbers of smaller kinds of animals were also introduced, including some 40 species of earthworms, 60 of spiders and mites, 12 of slugs and snails, and some 1100 species of insects, as well as around 1700 species of flowering plants; but virtually all of them (with a few prominent exceptions) have remained confined to the equally alien lowland farm and urban habitats provided by European man.[30] The modest behaviour of most species was, however, more than made up for by the excesses of the few that quickly reached spectacular numbers and spread over the whole country like a plague.

The first arrivals came in the late eighteenth century – sheep, goats and pigs on purpose, rats and cats as stowaways. The first of many introductions of the Australian brush-tailed possum, an appealing furry marsupial with a wide-eyed, cat-like face, came in 1837. The soft, woolly fur of the possum was seen as a valuable new crop that could be harvested from the untouched and otherwise unproductive forests. Domestic rabbits were liberated repeatedly from 1838 onwards; red deer and hares in 1851; fallow deer, and the first wild rabbits in 1864; hedgehogs in 1870; and sambar deer in 1875-76. Throughout the second half of the nineteenth century, there were repeated attempts to establish various other large game mammals for sport. Introductions ceased after about 1913, but by then seven different species of deer, plus chamois and Himalayan thar had settled in; only three failed to establish at least local populations.[31] It was not until much later that the unwelcome consequences of this policy began to show (p.110).

Predators of the early European period

By about halfway through the Polynesian occupation, all the most vulnerable native animals and birds had already gone, either to oblivion or to temporary havens on the offshore islands. The arrival of the early Europeans heralded a new invasion of predators with different skills, and those native species which had managed to come to terms with the Polynesians suddenly found themselves on the defensive again. Ironically, the predators of the Polynesian era were as much disadvantaged by the newcomers as were their former prey. Fortunately, we know more about the rapid environmental changes of the early European period than about those of the longer Polynesian era, largely because the early European explorers and settlers – with Cook, and later – included many careful observers of the natural scene. In their journals, sketches and specimens they recorded, for their patrons and thereby also for posterity, a priceless description of the islands and their wildlife during that first hundred years of European settlement. From them we can gain some idea of the radical changes in the numbers and distribution of native wildlife they witnessed in that short time.

Within the first few years of European colonization, the kiore and the true Maori dog began to decline in numbers, and both disappeared by the end of the nineteenth century. Some observers expected the Maoris themselves to do so too. It is usually assumed that the kiore was eliminated by competition from the introduced European rats. Charles Hursthouse wrote in 1857 that 'the kiore . . . is the only four-footed native creature which exists; and even this little king-quadruped of the country is fast disappearing before his alien congener from Norway'.[32] A handbook for colonists, written by Wakefield in 1848, states that there were two kinds of rats, the kiore and the 'large Norway rat, which exterminates the other race in its extensive migrations all over the islands'.[33] Dieffenbach, an early explorer who travelled through the North Island in 1839, mentioned that the natives distinguished between the Kiore Maori (indigenous rat)and the introduced Kiore Pakeha (strange rat) adding that 'on the former they fed very largely in former times; but now it has become so scarce, owing to the extermination carried on against it by the European rat, that I could never obtain one. . . . The natives never eat the latter. It is a favourite theme with them to speculate on their own extermination by the Europeans, in the same manner as the English rat has exterminated the indigenous rat.'[33]

However, the scarcity of the edible kiore, at least in the northern North Island where there was least beech forest and most rat-eating Maoris, may have pre-dated the arrival of the European rat. In 1769, Cook wrote, 'We saw no four-footed animals, either tame or wild, or signs of any, except dogs and rats, and these were very scarce, especially the latter.'[34] Von Hochstetter observed in 1858 that the kiore 'was so scarce already at the time of the arrival of the first Europeans, that a chief, on observing the large European rats on board one of the vessels, entreated the captain to let these rats run ashore, and thus enable the raising of some new and larger game.'[35] In 1869, W. L. Travers wrote to Thomson:

> It has been the fashion to assume that before the arrival of Europeans in this Colony, this creature was common, and to attribute its destruction to the European rat, and indeed, the natives have been credited with a proverb in relation to this point. It is not in effect impossible that the ultimate destruction of those which still existed when trade was first opened between Europeans and the natives . . . may have been hastened by the introduction of the European rat; but I am satisfied that before that time they had become very scarce, and indeed I have been told by gentlemen who have lived in the northern part of this island for upwards of forty years, that they never saw a specimen.[36]

The islands in the Hauraki Gulf are crucial refuges for native wildlife that can no longer survive on the mainland. Kiore were brought to Taranga, the main island in the Hen and Chickens group, by the Maoris; but fortunately they did not entirely destroy the value of Taranga as a refuge for several other sensitive native species, including the last natural population of the North Island saddleback. N.Z. Wildlife Service.

Whatever the cause of their disappearance, kiore persisted in the North Island only until the middle of the nineteenth century, and in most of the South until its end (a few still remain in Fiordland and South Westland). They are now found mainly on offshore islands, including at least 35 of the 146 'biologically significant' (i.e. not too much modified) islands for which information was available in 1976[37] – including Raoul, Hen and Chicken, Little Barrier, Kapiti and Stewart Islands.

Although the true Maori dog became extinct at the end of the nineteenth century, European-Maori dog hybrids appeared soon after settlement began, and many of both the introduced and crossed dogs went wild. Partly as a result of Cook's attempts to stock the islands with meat on the hoof, and partly due to strays from unfenced domestic flocks, the forests now contained feral pigs, goats and sheep. By the middle of the nineteenth century packs of wild curs were causing much trouble, chasing not only the feral pigs and sheep, but domestic ones as well. Contemporary accounts add that they fed on quail, ground larks, young ducks and kakapo.[38] But as settlement proceeded and the country became opened up, the wild dogs were gradually exterminated by the settlers. Even so, dogs in bush camps belonging to European explorers and surveyors killed many ground birds throughout the nineteenth century.

In this period at least, the Europeans did not bring in any predators on purpose. Their idea was to *increase* the stocks of native and introduced wildlife in their new home, and it was only because of their lack of understanding of how animal populations work that the results did not measure up to their intentions. The first predators to arrive, the ship-board cats and Norway rats, were uninvited; they simply lived 'wild' on the high seas, and in astonishing numbers. Melville described the rats on a whaling ship: 'They stood in their holes peering at you like old grandfathers in a doorway. Often they darted in upon us at meal times and nibbled our food. . . every chink and cranny swarmed with them; they did not live among you, but you among them.'[39] After long sea voyages round the world, many ships were brought close in or actually beached for maintenance, so there was no question of preventing the rats from landing, even if anyone had thought of it.

On Cook's first visit to New Zealand, he put in at Ship Cove in Queen Charlotte Sound in January 1770, and spent several days careening the *Endeavour* on the beach. On his second voyage he made straight for Dusky Bay, where the *Resolution* spent a month 'moored head and stern, so near the shore as to reach it with a brow or stage, which nature had in a manner prepared for us in a large tree, whose end or top reached our gunwale. Wood. . . was here so convenient that our yards were locked in the branches of the trees. . . .'[40] If Cook had been consciously trying to help the rats and cats on his ship to reach the shore, he could scarcely have made the access easier for them. These early four-footed pioneers were reinforced at more and more frequent intervals as the traffic in European ships increased over the succeeding years. The rat-infested ships of the sealers and whalers sailed all around the coasts of both main islands, and the rats no doubt lost no opportunity to get ashore. Conditions at the whaling stations made life easy for them; one contemporary description mentions 'In the water, huge sunken carcasses of whales; on the shore, great skulls, vertebrae, shoulder blades and ribs. . . the soil was impregnated with the smell of oil, and the air was loaded with the stench of decayed whale.'[41]

The Norway rat lives mainly on the ground – it is not good at climbing, although it

The Norway rat is the largest of the three species of rats brought to New Zealand by man. It has a thickset, heavy body and tail, a blunter nose and shorter ears than the other two. A quick way to distinguish between a Norway rat and the brown-backed phase of the ship rat is to turn the tail back along the body: in the former species the tail is shorter, in the latter, longer than the body. (In the kiore it can be either, but they are much smaller.) Cynthia Cass.

is an excellent swimmer. The larger, ground-nesting land and seabirds are the ones most vulnerable to Norway rats, rather than the smaller tree-nesting forest birds. According to I. A. E. Atkinson,[42] it was the Norway rat, not the ship rat, *R. rattus*, that was the common ship-board rat between 1710 and about 1830, so it was Norways that arrived with the earliest Europeans. They soon became very numerous, but did not remain confined for long to the coasts, nor did they leave with the ships that brought them. Norway rats rapidly spread over both islands; by 1840, early attempts to cultivate wheat on 30 acres at Riccarton, now one of the suburbs of Christchurch, had to be abandoned after less than a year, largely because of the scale of the losses due to rats;[c] and the early naturalists and explorers often remarked on the immense abundance of rats in the bush, especially during the 1870s and 1880s. Reading their words a hundred years later makes one realize just how much has changed in so short a time. Here are some examples.

The famous New Zealand ornithologist Walter Buller wrote in 1870: 'This cosmopolitan pest swarms through every part of the country, and nothing escapes its voracity. It is very abundant in all our woods, and the wonder rather is that any of our insessorial birds are able to rear their broods in safety. Species that nest in hollow trees, or in other situations accessible to the ravages of this little thief, are found to be decreasing, while other species whose nests are, as a rule, more favourably placed, continue to exist in undiminished numbers. . . .' Buller goes on to quote Rev. T. Chapman of Rotorua, who wrote to him 'some years ago' saying that 'Wild ducks were particularly numerous in this district on my arrival here: you saw them by the dozens – you hardly see them now by twos. I have no doubt we owe this to the Norway rat.'[43]

Sometime during the 1870s, G. M. Thomson visited Stewart Island, and was 'struck by the abundance of these animals in regions uninhabited and almost unvisited by man. One day the late Mr R. Paulin and I emerged from the bush on the south side of Thule in Paterson Inlet when the tide was low, exposing a wide stretch of beach nearly a mile long. We were very much surprised to find the whole beach alive with [Norway] rats which were feeding on the shellfish and stranded animals which the tide had left and exposed. As soon as they saw us they immediately ran for the shelter of the bush. They were literally in hundreds.'[44]

In 1877 Gillies described the hordes of rats in Otago during the early days of settlement:

. . . the rat (*Mus decumanus*)[d] was met with everywhere in great numbers. It was not confined to the neighbourhood of the settlements – Maori or whaling – but wherever you pitched your camp away in the wilderness, where never human foot before trod, there rats were found as abundant as near the settlers' homes. I remember distinctly on one occasion riding after a mob of cattle on a flat in the Taieri Plain near Otohiro in the year 1852, and seeing the rats running here and there in all directions from the horse's feet. When a new settler settled anywhere alone, the rats for a time were a perfect pest to him. They stole everything portable from him even to his candle-moulds, but after a time they became less and less numerous, and though they never disappeared wholly, yet nowhere in the country do rats swarm as they did in the early days. For years I was accustomed to camp out in new country miles away from any human being, but there were always plenty of rats.[e] On account of the dampness of the soil we used to make our fern or grass beds, if possible, on a bottom layer of dry branches, and we got so accustomed to the rats that we never felt inconvenienced by feeling them running below us through the branches or even over the top of us as we lay in bed. So tame were they that when the candle was lit in the tent they would come peering in at the door or under the curtain looking at you straight in the face with their earnest sharp gaze, and would only go when you shied something at them; they were not long in returning. On more than one occasion I have been present when men awoke with a rat lying right across their throat – we supposed for the sake of the warmth.[45]

Andreas Rcischek, an Austrian explorer and collector who came to New Zealand for two years in 1877 and stayed for 12, visited Fiordland in 1887, and graphically described the troubles he had with rats:

At Chalky Sound I had continual opportunity of observing the ravages of the brown rat (*Mus decumanus*), one of the great plagues of New Zealand. New Zealand, especially towards the sea, now swarms with these animals, introduced originally from European ships. They are a pest in the North Island, but round the sounds of the West Coast I found them more numerous still. . . . I regularly poisoned as many as I could. At night they kept me awake with their noise, knocking things down from the walls, gnawing at my stores and digging holes round the hut. They dug up the potatoes in the garden and dragged them away. . . . The tussock country near the Three Brothers swarmed with them, and they used to gnaw our books before our very eyes. While we were eating our supper by the fire, they would come along behind us and gnaw the bones we had thrown aside for Caesar . . . [one man asleep in camp] found a mob of them sitting round his head, gnawing his hair and beard, and shot out of bed as though a tarantula had stung him, got a stick and slew as many of his tormentors as he could. These rats are the great enemies of birds, and any bird living or breeding near the ground has but a small chance of existing.[f] They play havoc alike with eggs and young, and even attack the parent birds. . . . It took five months of shooting, poisoning and trapping before they showed signs of decreasing around camp.[46]

The coming of the Norway rats also spelt death to many native invertebrates, too, especially snails and large insects.

Towards the end of the early European period, it seems that, in the forests at least, the Norway rats in the North Island were in their turn replaced by a third species, the ship rat. (The other common name of this species, the black rat, is misleading, as less than a third of them are black; most are brown.) Whether the ship rats were later than Norways in arriving at the colony, or took longer to become established in the forests, we do not know. However, I. A. E. Atkinson has examined volumes of historical evidence,[42] from which he suggests that they became widespread in the North Island only after the 1860s, and in the South Island only after the 1890s. Hence the remarks of Douglas, an experienced explorer and surveyor in Westland, concerning the upper Waiho River (now in Westland National Park) in 1893: 'The Norwegian rat, which was no doubt responsible for some destruction, and which swarmed in the country at one time, is now becoming extinct from some cause or other, and the native and black rats are taking its place.'[47]

The ship rat is intermediate in size between the Norway rat and the kiore, but there are three distinct colour forms of ship rats, which freely interbreed with each other and have the same habits. One has a black back and grey belly; one has a brown back and grey belly; the third has a brown back and white belly. The ship rat is a natural climber, and spends much time in trees,[48] though it may also forage on the ground if

The ship rat probably came originally from South-East Asia, but its willingness to associate with man gave it a passport to the world – first along the mediaeval trade routes to Europe, and thence in European sailing ships around the globe. For a short period, roughly between 1710 and 1830, ship rats were apparently displaced from their maritime niche by Norway rats, but by 1850 at least they had regained it.[42] The reason for this interruption, if it happened, is not known, but in any event ship rats were well in command again by the time the serious traffic in immigrant ships to New Zealand began. Cynthia Cass.

other rats are absent. It usually nests above ground, in large untidy nests lodged among epiphytes, in a hollow branch, or an old bird's nest. With its long tail, it is very agile in trees and can run through the forest canopy like a squirrel. These characteristics make the ship rat a more effective predator of tree-nesting forest birds than the Norway rat or kiore, and certainly the records show that ship rats have been easily the most destructive of the three in historic times. That is not to say, of course, that the other two could not have been very damaging to the very different avifauna that existed earlier; we just do not have adequate observations, dating from then, in which the rat involved was unarguably identified. But we need not be in much doubt that things suddenly got much worse when ship rats joined them.

The last century has provided a number of documented examples of what happens when ship rats reach a previously untouched, relatively well-known community of birds. In New Zealand the best-known case is that of the Big South Cape Islands (off Stewart Island). Ship rats reached these islands in 1962 (or before) and developed a major irruption in 1964: four species of birds immediately became extinct.[49] Two were specialized endemics (Stead's bush wren and Stewart Island snipe) for which the islands were the *only* remaining habitat; and two were Stewart Island subspecies (the robin and fernbird, which still survive on Stewart Island). In addition, bellbirds, both red-crowned and yellow-crowned parakeets, and South Island saddlebacks were severely reduced; the bellbirds and parakeets later recovered somewhat, but the saddlebacks disappeared within the next five years. So

Big South Cape Island, off the coast of Stewart Island, had been an important predator-free refuge for native birds until 1964, when ship rats staged a massive irruption and wiped out the local populations of no less than five species of birds and one of the only two land mammals native to New Zealand, the short-tailed bat. Island refuges are extremely important for native species that cannot bear any predation; but as it is impossible to clear all rats from all commerical and private boats, the risk of this happening again remains (see p.156). N.Z. Wildlife Service.

did the last known colony of the Stewart Island short-tailed bat, which could have been a distinct subspecies. Without doubt, the Big South Cape incident can be regarded as the single most disastrous episode in the history of the wildlife of New Zealand, but it is certainly not a unique story. Elsewhere in the world, European rats (mainly the ship rat) have been involved with three quarters of all extinctions believed to have been caused by predators (Table 5). Prominent among them are the assaults on the birds of Lord Howe and the Hawaiian Islands by ship rats, described in Chapter 6.

Such stories incline one to favour Atkinson's suggestion that the rather late spread of ship rats through the main islands of New Zealand may account for the final disappearance of many bird species which had been able to hang on until then. This may be either because ship rats are more effective predators than the other two species, or because the effects they had were additional to any which the other two species had already exerted. Nowadays, mainland ship rats seldom eat birds,[50] but that is probably because few species remain on the mainland which are still vulnerable to rats.

Ship's cats must have been even more eager than ship's rats to try new hunting grounds ashore. Forster describes the *Resolution*'s cat in Dusky Bay in 1773: 'a sly cat on board, [she] had no sooner perceived so excellent an opportunity of obtaining delicious meals, than she regularly took a walk in the woods every morning and made great havoc among the little birds, that were not aware of such an insidious enemy'.[51] As the traffic in sailing ships increased, there is no doubt that many other ship's cats discovered the delights of hunting the gullible New Zealand birds – although at first, or at least until the population explosion staged by rabbits in the 1870s and onwards, feral cats usually stayed near the settlements.[52]

The first feral cats must have found living very easy. Besides the immense numbers of rats in the bush, there were also many native lizards and ground-feeding birds. The most famous (or infamous) cat of all was not even feral: it belonged to the unfortunate lighthouse keeper on Stephens Island (150 hectares) in Cook Strait, in 1894. This feline ornithologist simultaneously discovered and exterminated the entire world population of a hitherto unknown species, the Stephens Island wren. The only known specimens (about 22, of which 11 eventually reached museums) are those this cat brought in: only the lighthouse keeper ever saw one alive. Besides the wren, no less than 12 other species disappeared from Stephens Island, including the saddleback, kokako and native thrush. However, this amazing slaughter would have been achieved eventually anyway, since collectors no doubt helped the cat, and the lighthouse keepers cleared away almost all the island's forest for pasture.[53]

Cats have colonized 14 other unmodified, biologically important islands which are now reserves, and are implicated in the extinction of at least five other endemic species as well as in some 70 local extinctions. For example, in 1911 Guthrie Smith recorded nine species of birds on Herekopare Island (off Stewart Island) with a total population of some 400,000 birds. Cats colonized the island sometime between 1925 and 1931; by 1943 six species had gone and the other three were reduced to a few thousand.[54] On Little Barrier Island, cats eliminated the saddleback and kokako, and on Cuvier, the saddleback. This is by far the worst record of all the true carnivores accused of causing extinctions of birds – some 26 times worse than that for all three mustelids put together (Table 5). Cats were exterminated from Cuvier in 1961, from Herekopare Island in 1970, and from Little Barrier Island (p.157) in 1980.

In 1868 wild cats were very abundant on the Chatham Islands, and were destroying great numbers of indigenous birds. In 1911 on Mangere Island, one of the smaller islands in the Chatham group, one observer found a colony of tortoiseshell cats, the progeny of some liberated on the island to destroy the numerous rabbits there. He reported, 'I do not know whether they have succeeded in killing out the rabbits, but they have certainly exterminated the small native birds'[55] – including the Chatham Island fernbird and pigeon, and the black robin. The cats introduced to the Auckland Islands in 1820 no doubt had much to do with the extinction of the Auckland Island merganser by 1905. Cats helped to destroy the formerly teeming seabirds on Raoul Island in the Kermadec group, as well as the Kermadec parakeet; on Stewart Island, they threaten the last remaining breeding population of kakapo (p.148). Cats have gone from Mangere, but getting rid of them on islands as large as Chatham, Campbell, Auckland, Raoul or Stewart would be very difficult, if not impossible.

The activities of mainland populations of feral cats were less obvious, but they too could have caused local extinctions; in the early days, when the birds were more abundant and less well-known, these could conceivably have included unrecorded species. Charles Douglas remarked that wild cats must have killed many rock wrens in the high mountains, since cats were then (1899) to be found even up close to the snowline.[56] Douglas's friend and companion, A. P. Harper, wrote in 1895 that birds were becoming scarce in even the unexplored parts of Westland, due largely to the cats brought in by the gold diggers. 'The digger is very fond of his cat, and nearly always carries one with him; but in the past, when new "rushes" were frequent, he would go off at a moment's notice from his camp or hut, and if the cat was not at hand it was left behind, and naturally became wild. These have increased and multiplied enormously, and I have seen their tracks miles up unexplored valleys.'[57]

The fearless, curious behaviour of kokako and kakapo made them easy prey for a cat (p.102). In the only remaining place where cats and kakapo meet, in the southeast corner of Stewart Island, six of 118 cat droppings collected by the Wildlife Service contained kakapo feathers, and the remains of 14 kakapo eaten by cats have been found.[58] Otherwise the main prey of the Stewart Island cats are rats. The most intensively studied population of mainland cats in modern times, the ones living in the Orongorongo Valley near Wellington, now apparently eat rather few birds[59] – perhaps because there are few birds left there that feed or nest on the ground.

Besides those introduced species which were habitually predatory, we must not forget to mention others which helped, in less obvious ways, to make life more difficult for the native fauna. House mice became established in very early times, both in houses and in the bush, where they live quite independently of man, just as woodmice do in England. Their direct effect on birds is unknown, though not necessarily as small as one might expect: mice have been known to wreck the nests of birds, and to eat both eggs and dead nestlings.[60] But feral mice in the forest can be damaging indirectly, by helping to sustain other predators which do regularly eat birds. Outside the forests, rabbits have the same effect. There is one case in which the indirect role of additional mammalian prey in increasing predation on birds may

A feral cat photographed at night in the wild on DSIR Ecology Division's study area in the Orongorongo Valley, near Wellington. The small metal tag in the right ear shows that it is one of B. M. Fitzgerald's marked individuals. It was interrupted in the middle of eating its meal, a ship rat. G. D. Ward, DSIR Ecology Division.

have been the cause of an extinction. R. H. Taylor has pointed out that the ground-nesting Macquarie Island parakeets remained plentiful from the time of their discovery in 1810 until about 1880, although cats were present for almost all of that time, and feral dogs for the first 20 years. But after the successful introduction of rabbits in 1879, cats and wekas (ground birds native to New Zealand, but introduced to Macquarie Island) became much more abundant, and the parakeets became extinct by 1890.[61] The mice and ship rats, now also established there, arrived after the parakeets had gone.

Wild pigs, too, are not usually thought of as predators, but they would certainly never pass up the opportunity to crunch up a nest full of eggs, or a large native snail, that they might happen to come across lying on the ground. Pigs were among Captain Cook's first gifts to the native inhabitants of New Zealand in 1773, and more were released in 1777. By 1839, pigs were plentiful; Dieffenbach reported that 'the natives have great quantities of pigs, which have run wild, but are easily caught by dogs'.[62] Trade between the Maori and the early sealers and whalers consisted largely of pork and potatoes for muskets and gunpowder. The first effect of large numbers of wild pigs on the native fauna of the mainland was never described in detail, but no

Pighunting on the West Coast goldfields, a wood engraving from The Illustrated New Zealand Herald *of 1868. The pigs themselves look somewhat rotund and domestic compared with the real lean and savage 'Captain Cookers', but no doubt pigs of any kind would be damaging to ground-nesting birds. A note with the original states that 'there is every possibility that whilst so engaged many of [the diggers] have discovered fresh auriferous ground'. No doubt their dogs also discovered many ground birds and their eggs.* Alexander Turnbull Library.

doubt was comparable to that on the Auckland Islands. In 1909, Waite wrote: 'There can be small doubt that the introduction of pigs to the Auckland Islands [in 1807] has already resulted in considerable havoc among the ground-nesting birds, by destroying both eggs and young. Traces of pigs were very plentiful, not only their spoor but their rootings also being abundantly apparent. Native plants are also suffering, for we found whole patches turned over. . . .'[63]

However, it is probably stretching the point rather too far to include mice and pigs among the list of 'immigrant killers' brought to New Zealand, so we will not discuss them further.

Impact of predators on the native fauna

Not knowing that the land and fauna had already been radically changed by centuries of Polynesian occupation, Cook and his party probably assumed that they were witnessing wild nature in its original condition. They were greatly impressed by the variety and tameness of the birds, and never suspected that the most spectacular species of all were already extinct. Indeed, by then even the Maoris had only the haziest traditional memories of the moa,[64] and European scientists did not find out about them until 1843.[65] The birds that were present seemed marvellous enough to Cook, and in many places (not, of course, everywhere) abundant, easily caught and good eating. In 1769 he wrote: 'The country abounds with a great number of plants, and the woods with as great a variety of beautiful birds, many of them unknown to us.' Cook himself was more a navigator than an ornithologist, but nevertheless, when his ship was in Dusky Bay in 1773 he often remarked on the large numbers of 'ducks, woodhens and other wildfowl' he and his companions shot, as well as seals which 'are to be found in great numbers about this bay, on the small rocks and isles near the sea coast'. The party found coves where there were 'an immense number of woodhens', which 'as they cannot fly, they inhabit the skirts of the woods, and feed on the sea beach; and are so very tame, or foolish, as to stand and stare at us till we knocked

The weka is an inquisitive, aggressive bird, and the Maoris developed an absurdly simple but effective way of snaring them, as demonstrated here by William Fox's guide during his expedition to the Matukituki Valley in 1846. The lure is made of weka tail feathers, and the snare of plaited flax. Alexander Turnbull Library (detail only, by permission).

them down with a stick'. This habit, Cook correctly deduced, was the reason that, although the woodhens (wekas) 'are numerous enough here, [they] are so scarce in other parts, that I never saw but one. . . . The natives may have in a manner wholly destroyed them; they are a sort of rail, about the size, and a good deal like a common dung-hill hen, most of them are a dirty black or dark brown colour, and eat very well in a pie or fricassee.'[66]

Later European visitors continued in the same tradition. The moa-hunters had in their time already eliminated, or greatly reduced, a considerable number of bird species. But during the nineteenth century their place as bird-hunters was taken by an increasing number of European surveyors, explorers and collectors for museums. It was common practice in those days to set off for months in the bush with only a few staples such as flour and tea, relying on wild birds for meat. Douglas wrote, 'Years ago [in the 1860s] the . . . rivers in Southern Westland were celebrate for their ground birds, no prospector need carry meat with him, even a gun was unnecessary nothing was required but a dog, almost any mongrel would do'.[67] Pigeons were abundant and easily able to withstand modest cropping by Maori hunters equipped only with snares, but not the far more devastating effects of European firearms. Wakefield wrote in 1839: 'The wood-pigeons of this country are stupid . . . and, especially in these unfrequented parts, are not easily disturbed. . . . On the boughs of a small grove of trees, beneath which we lit our fire and disposed our beds and provisions, the pigeons settled in great numbers towards sunset. We had only to fire as quickly as the fowling pieces were loaded by the natives, hardly stirring from one

This remarkable photograph was taken on 24 October 1888 by Fred Muir, in the Clinton Valley, Fiordland. The man leaning on his gun in the centre is Quentin Mackinnon; he and the others were using the camp as the base for their expedition to find a route to Milford Sound – now the world-famous Milford Track. The four men and the dog killed large numbers of kakapo, blue duck and pigeons, and made pancakes with the blue duck eggs. At least four of the dead blue ducks are shown, and a live kakapo. National Museum.

position, the death of one bird not disturbing the equanimity of his companions on the same branch.'[68] The West Coast gold-diggers killed enormous numbers of pigeons; in Kumara about 1876 there was a man whose only work was to shoot them for the pot. He was known as 'Pigeon Tom', and he sold them to the meat-hungry diggers for a shilling a pair.[69]

Collectors for overseas museums were especially active in the later part of the nineteenth century, as scientific interest in the curious birds of New Zealand created a great demand for specimens. Though they may not have been numerous, the collectors had a special significance in that they concentrated on the rarest species, such as the huia. In 1874, a single expedition collected over 600 huia skins from the forested Tararua Ranges, north of Wellington. The local chiefs had prohibited huia hunting in the area for the previous seven years, in the hope of protecting them from being killed off.[70] In 1870 Potts wrote, about the little spotted kiwi, that

> the celebrity which attaches to this wingless genus is rapidly drawing down destruction upon it. No mercy is shown to it, and there is no exaggeration in stating that a regular trade is carried on in specimens of these birds, and the equally unfortunate Kakapo. Could not our paternal Government interfere on behalf of these interesting aborigines, for we believe there are those who would shoot the Cherubim for specimens, without the slightest remorse.[71]

One such person was certainly Andreas Reischek, who between 1882 and 1885 shot a total of 150 stitchbirds on Little Barrier Island, although stitchbirds were already extinct everywhere else in the country, and so rare on Little Barrier that he saw none on his first visit, and had to search for ten days of his second visit before he saw one.[72]

Many of the collectors, including some Maoris, worked only for the money, though a few felt that, since all the unique native birds were bound to become extinct anyway, the best thing to do was to ensure that good specimens reached as many museums as possible. As late as 1895, Buller was urging collectors to be sure to obtain complete, representative series of specimens of the fast-disappearing native birds, for each of the existing museums, 'before it is too late'.[73] All the same, Michael King's comment about Reischek (which could easily be applied to other ornithologists of the day, including Buller) hits the nail right on the head: 'While it is true that Reischek has to be observed against what were accepted standards at the time, the scale of his shooting, and the fact that, for example, he later [in 1887] used kokako for soup, tends to diminish the admiration of a twentieth-century observer.'[74]

The reasons for this behaviour are exactly the same as those that led to the destruction of the seals, the whales and the kauri – and, in our own time, the lowland forests: these species could be exploited for private gain at a cost which did not reflect society's loss, and the reduction in their numbers was a cause, rather than a consequence, of the over-exploitation.[75] Reischek's own attitude illustrates this approach rather well; he himself believed that the stitchbirds on Little Barrier Island would soon be exterminated by the cats, and he obviously intended to make as much money as possible by supplying specimens to museums while he still could. Naturally, the rarer the species, the higher the price, and the greater the effort he was prepared to make.

During the early European period seven species or subspecies of mainland birds

Andreas Reischek, in middle age, posing for a studio portrait in memory of his New Zealand expeditions. The axe and hatchet were necessary to make headway and camps in the bush: the gun he used to all too good effect, especially on the rarest species he sought most ardently for his collections. Alexander Turnbull Library.

suffered extinction or irreversible declines in density (Table 2). The list of birds affected in this period is somewhat arbitrary, since some of them may already have begun to decline during the Polynesian era; but, so far as we can tell, these seven species or subspecies were still abundant and widely distributed before the Europeans arrived. Of the seven, four became totally extinct: the New Zealand quail on both islands, the North Island laughing owl, the North Island thrush, and the North Island bush wren. Two, the stitchbird and North Island saddleback, became extinct on the mainland, but still survive on offshore islands. A seventh, the huia, lingered on until about 1907, but declined steeply towards the end of this period. The stitchbird and the huia were confined to the North Island. Also in the North Island, the bellbird and North Island robin became temporarily very scarce in the 1860s and 1870s, though both later recovered. What immediately strikes one about this list is that most are smaller birds than the ones that had disappeared in Polynesian times, and that most of them lived in the North Island. In addition, the three species that were already declining before the arrival of the Europeans – the takahe, kakapo and little spotted kiwi – declined further during this period, and all were extinct on the North Island by the end of it. Of course, it is unlikely that any were affected by only one factor, but for some, we can guess what was the final straw.

The New Zealand quail is (or rather, was) a small insectivorous ground-dwelling bird, closely related to the Australian stubble quail,[g] which suggests that it arrived in New Zealand relatively recently (p.29), probably between 5000 and 10,000 years ago. In the North Island, quail never did live in the extensive forests still standing in 1840, only in the rather limited open areas, which were the first to be occupied by the settlers and their rats and cats: it was gone by 1869. In the wide open plains of the South Island, the quail were once extremely abundant, 'round my station in thousands', as one runholder said. Near Nelson, one man shot 59 birds in a few hours in 1848, then prudently forbade shooting for a year to build up the stocks. When he

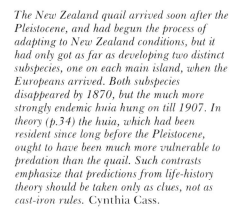

The New Zealand quail arrived soon after the Pleistocene, and had begun the process of adapting to New Zealand conditions, but it had only got as far as developing two distinct subspecies, one on each main island, when the Europeans arrived. Both subspecies disappeared by 1870, but the much more strongly endemic huia hung on till 1907. In theory (p.34) the huia, which had been resident since long before the Pleistocene, ought to have been much more vulnerable to predation than the quail. Such contrasts emphasize that predictions from life-history theory should be taken only as clues, not as cast-iron rules. Cynthia Cass.

went back in 1850, expecting some good sport, there were none left.[76] Fires (destroying cover, insects and birds alike), poisoned grain spread for rabbit control, or rats, cats and feral dogs, would all have contributed to the near-inevitable extinction of the South Island quail; but a virulent epidemic would explain its suddenness. It was gone by 1870. Some other species suffered temporary setbacks at about the same time – for example, the bellbird. Bellbirds are now widespread and relatively common in lowland podocarp forests, but in 1877 Buller believed them to be on the brink of extinction.[77] There were other declines that were also very sudden, but permanent: for example, in parakeets in the 1890s, and in wekas and brown teal in the 1930s.[78]

The explanation offered for the most sudden of these declines is that native birds must be especially susceptible to unfamiliar avian diseases imported along with introduced birds. Unfortunately there is no real evidence for or against the idea, though it is quite plausible. But the effects of introduced parasites and germs on native birds in the last century is unknown, and their effects now are apparently minor and local. Moreover, there are no grounds to suppose that the native birds were necessarily entirely free from avian diseases even before the era of European introductions, since there was always the chance of pathogens arriving with migrant birds.

The huia had a special significance for the Maoris, who greatly prized the large black tail feathers, tipped with white.[79] Huia feathers were used as a badge of rank, and passed from tribe to tribe by means of barter, and so reached even the far north and the South Island, where the bird itself was never found. In European times, the huia was decimated by avid collecting for museum specimens; it lingered on to the turn of the century, and could have been finished off by mustelids, 'those furred imported devils of the animal kingdom',[80] especially in areas not visited by collectors;

but its decline was already well advanced, perhaps irreversible, before mustelids became widespread. The thrush, bush wren, stitchbird, saddleback, bellbird and robin were all ground-dwelling birds, unusually vulnerable to rats, and their declines coincided with the spread of the ship rat in the North Island. G. R. Williams summed up the position well: 'If introduced mammals, and not disease, were responsible for the disappearance or increasing rarity of a number of mainland species during the middle of the nineteenth century, or even earlier, then only *R. rattus* and *R. norvegicus* would be likely to have been sufficiently numerous and widespread to have been responsible.'[81]

Conclusion

When the Europeans first arrived, the native animals and birds that were still living on the mainland were only those that had managed to find ways of co-existing with the Maoris and the kiore, or else had found a large enough refuge in the still very extensive forests that remained. However, what precarious truce had been achieved in those years was only temporary. With the rising tide of European invaders that followed Cook, and with their various animal companions, the extent and pace of disturbance quickly accelerated. The impact of the new predators on the forest birds of the North Island in the nineteenth century was comparable, though on a smaller scale, with that of the early Polynesian hunters on the large game birds of the South Island in the twelfth and thirteenth centuries, and for the same reason. The slow-breeding and flightless moa, rails and waterfowl had no defence against the spears and axes of the moa-hunters; the slow-breeding and weak-flying huia, saddleback, and native thrush had no defence against rats, cats, dogs and European firearms. As in the Polynesian period, the rapid disappearance of these birds was not primarily due to the prowess of the invaders, but to the naked vulnerability of the birds themselves.

There are alternative explanations, of course, but none of comparable scale and scope. Forest clearing by Europeans in the North Island was only just getting into its stride by 1884, and in any event, the diminution of birds was evident even in the huge, untouched forests of the inland North Island;[82] repeated burning of the tussock by sheep-runholders in the South Island was certainly a major cause of the disappearance of quail and native snails, but did not affect forest birds; introduced avian diseases could be carried into the forest only to the extent that the exotic birds could gain a foothold, which at this time was largely limited to the peripheral areas already disturbed by fire and browsing stock; the introduced deer and possums did not reach their devastating peaks of abundance in the forests until well into the twentieth century. When habitat and food supplies have been destroyed, birds will disappear whether or not predators help to speed them on their way: but in the early European period, birds were disappearing from huge tracts of largely unmodified forest. The indications are even stronger than for Polynesian times that the birds that suffered most in the first century of European occupation were felled by the immigrant killers of the time – especially rats and cats, which long preceded European settlement and clearing into the remotest areas. They were certainly lethal enough, but even they were not the end of it all: there was worse to come, and in 1884 the latest and deadliest onslaught began.

A professional rabbiter with his horsé, gun and eight dogs in about 1910. Despite his determined expression, the direct predation by man on rabbits never provided the final solution, partly because no rabbiter wanted to be so efficient as to put himself out of a job, and partly because the more rabbits are killed by direct methods that do not alter the habitat, the better conditions are for the breeding and survival of the remainder.
Canterbury Museum.

4

THE LATER EUROPEAN PERIOD
1884–1984

Drastic measures

By the mid-nineteenth century, the pastoral economy of the South Island was well established. The terrible labour of founding a new colony from nothing – building houses, yards and woolsheds entirely from scratch, finding ways of handling sheep without fences and carting enormous bales of wool over appalling roads – were beginning to pay off. The sheepruns were prosperous, homes were comfortable; the civilizing influences of arts and science had time to take root. All this had been achieved in a mere 20 or 30 years; with the northern land wars over, the worst hurdle seemed to be past. But there was trouble brewing of a different kind, and by the 1870s it was reaching serious proportions.

The lack of native game to shoot at had been an inconvenience for the early settlers, but not for long. The Acclimatization Societies had set out to enrich impoverished New Zealand with the choicest products of the animal kingdom, as they hastened to repair nature's omissions among the native wildlife with what they regarded as the very best possible imports. Domestic rabbits were among the first animals brought in to supply the settlers' cooking pots; although some escaped, or were turned out, few survived. But real wild rabbits were introduced in Southland in 1864,[1] and, after several false starts, they eventually got established and began to behave in the proverbial way. Before long they were becoming a serious nuisance in the south, and were spreading northwards in an inexorable army.

By the 1870s, runholders were becoming alarmed by the economic loss and soil erosion caused by rabbits. To quote only two examples, the flock at Moa Flat Station in Otago was reduced from 120,000 to 45,000 sheep; that on Castle Rock Station in Southland, from 50,000 to 20,000, while 300,000 rabbits were killed in one year.[2] The fine natural grasses on which sheep and cattle grazed were almost totally destroyed. Sheep died from starvation by the hundreds of thousands, and it is no exaggeration to say that the majority of the high-country pastures were ruined. From 1878 onwards, immense areas of grazing land were abandoned, as the owners gave up the unequal struggle with the rabbits. Millions were killed, without the slightest effect: they passed over the country like a tussock fire.

The afflicted runholders reasoned that, since rabbits did not reach such huge numbers in their original home, their rapid multiplication in New Zealand must be due to the absence of their natural predators – the foxes, stoats, polecats, hawks and owls of the Northern Hemisphere. The obvious remedy would be to restore the 'balance of nature' (much talked of, but little understood, then or now) and things would return to normal. The farmers demanded that the government bring in the 'natural enemies' of the rabbit – starting with ferrets.

The ferret is the domesticated version of the wild polecat. It has been bred in captivity at least since Roman times, but we do not know whether the original stock were European polecats (Mustela putorius) *or Asiatic steppe polecats* (Mustela eversmanni). *To save confusion, tame ferrets, and their feral descendants in New Zealand, are usually treated as a separate species,* Mustela furo. *Ferrets were supposed to control the rabbit pest in the 1880s, but in reality, the rabbits controlled the distribution and numbers of ferrets, as they still do.* Cynthia Cass.

The ferret is a domesticated version of the wild polecat, and is a renowned rabbit-hunter. Ferrets could easily be obtained from gamekeepers in England and from rabbiters in Australia; in fact, some were already present in New Zealand before the debate about rabbit predators had even begun. Five ferrets arrived in Canterbury in 1867, but were kept in captivity, more or less as pets. But releasing them in the wild would be a quite different matter. Ornithologists and some of the Acclimatization Societies argued strongly against the idea – although, as the societies had helped to introduce the rabbits, they were not in a strong position.[a] But farmers saw it as a simple, cheap solution to an unforseen check on development. As time went on, and things began to get desperate, they began to see it as the last possible chance to save the entire future of pastoral farming in New Zealand. The issues were clear, and well summarised by Buller at the meeting of the Wellington Philosophical Society on 9th December, 1876.

Buller reported that 'the Legislature having rejected the proposed measure for prohibiting the introduction of polecats and other noxious animals into this colony, nothing now remains for us but to sound the note of warning before it is too late, and, by directing public opinion to the subject, to mitigate the danger of our being overrun with one of the worst of predaceous vermin'. He then quoted a letter from Professor Newton of Cambridge, strongly advising against the idea, since in England 'the ·polecat . . . is the most detested beast we have. . . . In New Zealand (if introduced) it will quickly become master of the situation . . . and then goodbye at once and for ever to all your brevipennate [flightless] birds, as well as to many other of your native species, which of course have no instincts whereby they may escape

Sir Walter Lowry Buller in 1905. Buller was tireless in his opposition to the proposed introduction of mustelids into New Zealand, and continued to denounce the idea for years after it was too late. A hundred years later we can certainly agree that it was the wrong thing to do, but with the advantage of a longer perspective over the history of New Zealand's birds, we can see that Buller's fears were exaggerated: we cannot at present identify any native species which has been exterminated by any of the mustelids (p.106). Alexander Turnbull Library.

from such bloodthirsty enemies – to say nothing of pheasants and the like, which you have been introducing at so great a cost, and your poultry. . . . I trust that, among you all, you will be able to keep off this threatened scourge. Colonists in general have not been slow to hinder unacceptable importations from the Mother Country, as witness the historic tea-chests at Boston (USA), and Australian convicts . . . there cannot be a doubt of how you should behave when you have a shipload of known murderers to be let loose on your peaceful shores. . . .'[3]

Buller rightly anticipated that 'the answer to this will naturally be that the bloodthirsty invader will do good service in ridding the country of rabbits; but the grave question to be considered is whether, in the attempt to put down one evil, you are not permitting a larger one to grow up in its place. The polecat once established in this country, it would be almost impossible to extirpate it; and the disturbance to the existing conditions of animal life, by such an introduction as that proposed, would be incalculable. It will, no doubt, be argued on the other side, that sheep are of more practical account to the colony than kiwis and wekas, and that, from the sheepfarmers' point of view, anything almost is preferable to the "rabbit nuisance". But the real point raised by Professor Newton and deserving of earnest consideration, is whether the object in view cannot be attained by other equally efficacious means and without the introduction of the pestilent polecat.' Through the somewhat pompous language of the nineteenth century we can recognize some of the tensions between development and preservation, profit and aesthetics, haste and deliberation, which we still face today. But then, as now, insufficient time could be allowed for earnest consideration while commercial interests had the power and

Rabbits prefer short, dry grass and light sandy soils. Paddocks and tussock grasslands already overstocked with sheep have many bare patches favourable to rabbits, which they tend to concentrate on, ignoring tall rank growth. In the dry eastern grasslands of both main islands, patches of short grass covered with rabbit droppings must have been a common and depressing sight for farmers of the 1870s. Naturally they wanted a quick, cheap solution, but it was many years before research was able to show how bad farm management aggravated the problem, and why killing individual rabbits with guns, dogs, traps or predators was not the answer. W. E. Howard, DSIR Ecology Division.

the influence to bring up the big guns; and so the protests were overruled.

The earliest known deliberate release was of five ferrets carried in backpacks to the valley of the Conway River in 1879, but this seems to have been an isolated instance, and there is no record of the result. Ferrets were first released in numbers in 1882, and in 1883 the Chief Rabbit Inspector, a Mr Bailey, recommended the additional introduction of stoats and weasels into the back country for rabbit control. The first shipment, imported privately by a Mr Rich of Palmerston, arrived in 1884, and for the next ten years hundreds of stoats and weasels, and thousands of ferrets, were turned out, both by government agents and from private shipments, on the farmlands worst affected by rabbits. Most of the stoats and weasels probably came from British stock; there were about 17,000 gamekeepers employed on 'vermin control' on the great game estates in Britain at the time,[4] who were no doubt delighted to be paid twice for catching these relatively abundant small predators on their beats. The New Zealand Department of Agriculture bred ferrets for release until 1897, and private individuals continued until 1912, producing about 300 a year.

Although it was already too late, the protests continued. H. B. Martin told the Nelson Philosophical Society in 1884 that:

> The introduction of these beasts of prey to destroy the rabbit is unnecessary, for poisoning with phosphorised corn succeeds well . . . while tuberculosis (which has recently broken out among the rabbits in Otago) will probably destroy them more thoroughly than any other means would. . . . Having no natural enemies here and their furs being of very inferior

quality in this climate, there would be no adequate check on them, and they would therefore increase and spread as the rabbit has done. In Canada and other northern regions the weasels are killed in great numbers for their furs, and are also preyed on by larger beasts of prey, while in more settled districts their ravages among game and poultry cause them very generally to be destroyed, yet with all this they are in no danger of extinction even where most persecuted . . . and in England the stoat and weasel are so common, though freely destroyed, that it would seem impossible to exterminate them[b]. . . . They have no marked preference for any one species of animal, but habitually live on birds and small mammals, so that being very lithe and agile, and for the most part active climbers and bold swimmers, no species of bird would escape their ravages. . . . The stoat, for example, can climb any tree, and is so light and active, that any branch is accessible to it that will bear the weight of nest and eggs. . . . The importation of these beasts should therefore be stopped, and those at liberty destroyed, at whatever cost; if this is done without delay, I do not think it is now too late to extirpate them.[6]

And Reischek told the Auckland Institute in October 1885:

. . . if stoats, ferrets, weasels . . . are turned out to destroy rabbits, it will be difficult to protect the birds, as these creatures destroy them, especially ground birds such as kiwis, kakapos, wrens . . . in Austria we destroy these animals at every opportunity. They are very cunning, and will not take poison while they can get live prey. Rabbits are much easier destroyed by shooting, netting, or bagging with [?tame] ferrets when the land becomes more closely settled.[7]

But these appeals fell on deaf ears, and the imports continued.

In 1885 and 1886 alone, 214 stoats, 592 weasels and thousands of ferrets were liberated. The earliest release areas were Lake Wanaka, the Makaroa and Wilkin

The stoat is an agile climber, and the rough surface of this punga (tree-fern) trunk would offer it ideal footing. Stoats can climb large trees to the full height of the forest canopy, running fearlessly along the branches and down again head first. A. Brandon, Taranaki Daily News Co. Ltd.

Valleys, Lake Wakatipu, part of Southland, Ashburton, Lake Ohau, Waitaki, Marlborough (3000 ferrets in 1884) and West Wairarapa; they were selected for the abundance of their rabbits. Most of the shipments were arranged through the New Zealand Agent-General in London, and at no small expense: a pair of weasels would cost £5, and ferrets were £3. 2s.7d. a head. About 3000 stoats and weasels were sent from Lincolnshire in 1885. Eventually, British farmers began to protest, and at the inquiry into the 1882 vole plague in Scotland, at least one witness blamed the sudden increase of rodents on the wholesale collection of weasels for export to the colonies. He suggested re-importing them![8] The efforts of the Agent-General were supplemented by private associations of New Zealand landowners, formed in 1886 to collect money for the provision and liberation of mustelids. They sincerely believed that introduced carnivores would be the best weapon against the rabbit pest.

It soon became clear, however, that the new arrivals were not having the expected dramatic effect. In the early years, a reduction of rabbits south of the Clarence River, in eastern Marlborough, was attributed to mustelids, but it is not now possible to say whether this is true, or whether the rabbits were simply going through one of their natural temporary declines. At Mount Somers, in Canterbury, the rabbits seemed to increase a hundred per cent or more, and some weasel nests found contained only skylark feathers; Mr Peters, the runholder, reckoned that the wild cats already present at Mount Somers were far more effective in keeping down rabbits than were stoats and weasels, and he estimated that a cat would kill more rabbits in a month than a stoat or weasel would in six months. This belief (though probably wrong) was quite general, and farmers often sent men into the towns to buy up cats and bring them back for liberation on their rabbit-sick pastures. But however many rabbits the cats killed, the huge rabbit populations remained unaffected – especially in places where traps were used, in which cats and mustelids were often caught accidentally.[9]

Within six years came the first reports of weasels and stoats spreading into forests far from their release sites, and of drastic declines in native birds in the forests of Westland and Fiordland. In 1892, Walsh remarked that 'stoats and weasels, from which so much was expected, have not only failed to accomplish the object desired, but are already, in the destruction of native birds, and in their depredations of the fowl yard, proving themselves an intolerable nuisance'.[10] Buller believed that his fears had been amply realized, and his cries of I-told-you-so reached new heights of indignation. In 1894 he raged, 'I regard this act in the light of a crime. The vermin that every farmer in the Old Country was trying to extirpate as an unmitigated evil our wise Government bought up by the hundred and imported into this country, in the vain hope that these "carnivorous beasts" would change their habits and take to a rabbit diet, to the exclusion of everything else!. . . I raised my voice in protest at so insane a policy . . . but all to no purpose. The imported animals were turned loose north and south, and have now become so firmly acclimatized in a country where the conditions of life are so favourable to their existence that no power on earth will ever dislodge them'.[11] It must have been small comfort to him to see his gloomy prognosis proved so sadly right.

By the turn of the century, even their former supporters could see that the passage-paid overseas experts were no match for the rabbits. The Acclimatization Societies – at least, those that had not too publicly supported the now-discredited idea of bringing in predators – complained bitterly that the mustelids and the poisoned grain, both intended to control rabbits, were instead 'responsible for the disappearance of our native and extinction of our game birds'[12] – although the early liberations of game birds had been successful. H. Guthrie-Smith was normally a careful and objective observer of the succession of rapid changes in wildlife populations he witnessed in his lifetime; but he was not at all objective about the Wairarapa farmers who had so eagerly supported the introduction of first the rabbits, and then the mustelids. He denounced them as having attempted to 'correct a blunder by a crime', and wished them all at the bottom of the sea.[13] He reported that the first weasels reached Tutira, his sheepstation in Hawkes Bay, in 1902, spreading north from the Wairarapa *ahead* of the rabbits. For three years they were very numerous; he mentions accounts of weasels or stoats attacking lambs and even children, men and horses, which he checked on, and thought some at least to be true.

On the other hand, G. M. Thomson, fairminded as ever, pointed out that the account was not entirely negative: '. . . one direct benefit which stoats and weasels confer is the wholesale destruction of rats and mice which they cause . . . it may be that rats are more destructive to eggs and to young birds than even stoats and weasels . . . twenty years ago rats were a perfect curse about the homesteads, destroying harness, sheepskins, grain and food, but since the weasels appeared the rats have absolutely gone'.[14] Thomson also considered that the evidence concerning the destruction of the native avifauna was at that time (1921) inconclusive. He pointed out that mustelids had spread into some areas without rabbits, where they presumably lived only on rats and birds, and yet even there, native birds survived in abundance. In the North Island at least this assessment could well have been correct, because by that time all the most vulnerable North Island species were long gone, and the ones remaining would no doubt be relieved to see the new arrivals reducing the plagues of rats.

The realization of what had been done must have been awful to New Zealand naturalists, and it caused terrible arguments and recriminations. As Thomson wrote, 'Nothing in connection with the naturalization of wild animals into New Zealand has caused so much heartburning and controversy as the introduction of these bloodthirsty creatures.'[15] The government changed its policy, but it was too late. In 1903 an amendment to the Rabbit Act removed the law's protection from the so-called 'natural enemies of the rabbit' which had fallen down on the job. When the Acclimatization Societies began to offer bounties for mustelid tails, they spent £3,738 in the years 1939-48 in the North Island (37,384 mustelids at two shillings a tail) but with as little effect on mustelid populations as the mustelids had originally had on the rabbits. Bounty payments ended in 1950 because, as in other countries where they had been tried, they only cropped, not controlled, the target populations. In the South Island, ferret and stoat furs were valuable – top-grade skins fetched 16 to 18 shillings each; but even the harvesting of 140,000 ferrets and 51,000 stoats in the five years 1944-48 made no apparent impression on their numbers.[16] The problems and expense incurred by the original malady were now just about matched by those stemming from the supposed cure: both rabbits and mustelids defied all attempts to control them until the 1950s and 60s.

The sad thing about it all is that the dawning realization by the government of the need to protect the endemic native fauna was already growing rapidly at the very time that the introductions were going on, but it did not gain enough strength – or at least, enough to make it a force to be reckoned with – in time to prevent the introductions altogether. As H. G. Wells remarked, 'Human history becomes more and more a race between education and catastrophe' –[17] and in this case, catastrophe won. However, we did at least learn something from the experience. Perhaps because the consequences of the predator-introduction policy were becoming all too clear in the 1890s, the government was persuaded to secure the permanent protection of three important reserves – Little Barrier, Resolution and Kapiti Islands – by 1895. It is true that they were not commercially attractive places, but nevertheless, few other governments in the world at that time were interested in conservation at all.

On the mainland, the importations continued even after 1895, though not without comment from the indefatigable Buller: 'I regard with extreme satisfaction this gradual awakening to the fact that we have animal and vegetable forms of life, indigenous to the country, which ought to be protected and cherished. . . . Forest reserves . . . are being defined and proclaimed; and the law is being invoked for the protection, one after another, of our rare species of birds. The only danger to be apprehended now is that by continuing the insane policy of introducing predatory animals, such as stoats, ferrets and weasels, in the vain hope of suppressing the rabbit nuisance, the good that is being accomplished may be to a great extent counterbalanced. . . . It seems to me that the only chance of arresting this deplorable evil is by directing public opinion against it. . . . Unfortunately, most people are indifferent about it, and the government yields to the clamour of a few faddists whose one idea is to exterminate the rabbits at any cost to the country.'[18] The situation sounds remarkably familiar.

Nowadays we know rather more about the ecology of animals than the early pioneers did, and we may wonder at the naivety of those who saw the introduction of mustelids as the solution to the rabbit problem. But do not forget that in those days it was a simple matter of survival; there was an economic depression on, and those farmers were struggling for their lives. There was no comparison, for most of them, between the value of the native wildlife and that of their property, considering the grinding physical labour that it cost to run the average nineteenth-century farm. The forces of economics are more compelling than those of conservation. We should not blame Mr Bailey and his colleagues; the introductions probably could not have been avoided, given the economic pressures of the times. Neither should we feel at all superior, since we are still capable of making equally unwise decisions. With hindsight we can all agree with Melland that it was 'incredible folly' for the government to turn out ferrets on the west shore of Lake Manapouri in the 1880s,[19] where kakapo, then still common, are now extinct; but how will people a hundred years from now judge us, who in the 1980s failed to prohibit the destruction of the last few lowland native forests and their inhabitants, merely to postpone, for a few more years, the inevitable closure of a few uneconomic sawmills?

Land ownership and deforestation

Long before the beginning of the later European period in 1884, the plagues of rabbits were having a dramatic effect on the already faltering native pastures: and they were followed by more and worse environmental shocks in the next hundred years. The transformation of the rural landscape by the imposition of European agriculture was already under way, but what had been achieved so far was nothing compared with what was to follow.

At first, many settlers held only leasehold rights over their land, which still technically belonged to Maori owners, but this arrangement soon became inadequate. Men who had put the best twenty years of their lives into clearing a bushclad section felt that every penny of its increased value should belong to them; besides, many of the colonists had been attracted to New Zealand in the first place by the prospect of owning their own land, becoming their own masters and being able to pass the fruit of their labours on to their sons.[20] So the North Island leaseholders put great pressure on the government to allow them to purchase their leasehold properties at the original valuation; as the amount of land owned and cleared by Europeans increased, so that owned by Maoris (much still in forest) declined.

In 1840, Maoris had owned almost the whole country (26 million hectares): they did not occupy it all, but between them the tribes laid claim to most of it.[21] The parts they did not live in they visited periodically to substantiate their claim and to check on the boundaries, marked by natural features or boundary stones. This visiting was necessary because, in the Maori system of land tenure, ownership of land claimed by right of discovery, conquest or inheritance had to be confirmed by occupation, or the claim could be lost. Their concept of land tenure was communal (i.e. vested in the whole tribe), but participation in it was intensely personal, and bound up with social standing and the right to speak at tribal gatherings. They did not think of the land on which they had such rights as a commodity, to be bought or sold, but as *turangawaewae* (literally, a standing-place for the feet). 'If we lose our land,' they would say, 'we lose our *turangawaewae*.'[22]

Europeans could not understand this attitude, and by the 1890s they were sufficiently dominant, both in numbers and in economic and political power, to ignore it; for their part, the remnants of the originally vigorous population of Maoris that were left were too demoralized to put up an effective defence of their ancient way of life. By 1892, Maoris retained only one sixth of the country (4.4 million hectares).[23] In a mere 52 years since the 'official' start of European colonization – which had originally guaranteed, in the Treaty of Waitangi, to confirm the rights of the chiefs to hold their lands – over 20 million hectares had passed into European or Crown ownership. These 'alienated' lands, comprising very large areas of the North Island, and almost all of the South Island, included all the most productive agricultural soils. The Maori lands remaining were mostly rugged, remote and bushclad. They were left in Maori hands largely because Europeans thought them useless. In fact, of course, they were of immense value, at least to wildlife: until the 1890s, before the government began to think seriously about forest reserves, the Maori lands were important refuges for the beleagured native fauna. New Zealand's Pakeha conservationists are rightly proud of the fact that their first National Park, Tongariro, was set aside as early as 1887, only 15 years after the very first National Park in the world, Yellowstone, was established in North America. They often forget to mention that this was possible only because the mountains protected by the park

The sacred mountain of Ruapehu, given to the Government by its Maori owners in 1887 on condition that it be protected as a National Park. The gift was accepted mainly because of its outstanding scenic beauty; as a reserve for threatened wildlife it was less important than other, less striking Maori lands which have since been sold and cleared. Perhaps if the Maoris had been strong enough to retain more of their tribal lands through the century after 1850, there might be more pristine lowland habitat to protect now that we realize the need for habitat conservation for its own sake. Author.

were sacred to the local Maoris, and their paramount chief, Te Heu Heu Tukino, saw that the only way to defend them from the greedy settlers was to present them to the government, on condition that they were protected from all development.[24]

The alienation of Maori land was very rapid around the turn of the century, when the Maori population was at its lowest ebb. The Darwinian prognosis that 'native' races confronted with European colonization were doomed to extinction seemed to be confirmed; the Maori would share the fate of the Tasmanian Aboriginal. It seems shocking to us now, but then it was obvious and rational that most Europeans should accept this prognosis with equanimity and, with clear consciences, continue to acquire Maori land and destroy the native vegetation that stood in the way of increased production of wool, meat, butter and cheese. Between 1896 and 1921 more than half of the Maoris' remaining landholdings – 2.4 million hectares – were sold.[25] The Maoris, who had been the dominant culture and the main agricultural producers of the 1840s and 1850s, were now reduced to appalling subsistence conditions, often with insufficient medical attention or sanitation, their social structure disintegrating or unable to cope with the changes in their lives brought in by European dominance and indifference, their traditional remedies ineffective against European diseases. True, there was some degree of response to European culture, interpreted to mean that the Maori was becoming 'civilized'; but in the main, the Maori and Pakeha cultures co-existed, sometimes as it were in the same bed, more often in separate beds in the same house. But the Pakeha had got hold of the house.[26] By 1974, the Maori lands totalled a mere 1.4 million hectares (not counting freehold land held by Maoris under individual, as opposed to communal or tribal, title).

For the purposes of this book, I have dated the later European period as beginning with the first introduction of stoats in 1884. But their arrival more or less coincided with another important development in the technology of farming, which stimulated yet further rapid increases in the extent of deforestation and 'improvement' of virgin land. It was the introduction, in 1882, of refrigeration in the dairy factories and freezing works processing huge amounts of butter, cheese and meat for export, and in the ships carrying these products to Britain. This impetus to increase productivity for an assured export market, coupled with the opening of vast areas of the North Island for settlement after the end of the land wars, led to boom times for land- and farm-stock agents, assisted by a steady drift of settlers from the South to the North Island from the 1880s onwards. Previously, the terrible sweat of carving a small farm out of the thick North Island bush had been worthwhile only within horse-and-cart distance (about five kilometres) from the nearest dairy factory. Now, settlers began to clear land that had, until then, been too far away. The number of dairy farmers, 'minute' in 1891, increased to 5000 in 1901, and three times that – a third of all farmers – in 1911. By the beginning of the First World War, they had taken over most of the accessible and naturally fertile low-lying land, cleared, drained, and fenced it, sown it with European pasture grasses, and equipped it with cows, roads and dozens of small dairy factories. Above the enclosed fields, safe only for the moment, the steeper hills retained their forest and their isolation.

After the war, New Zealand embarked on an ambitious policy of repaying the returned war heroes with pieces of the land that they had fought for. Under the 'commandeer system', Britain gave New Zealand's staple exports – wool, meat and dairy products – a guaranteed market and high prices in Britain in return for a monopoly of access to these primary products during and after the war. So thousands of soldier-settlers took up sections they had never seen, mortgaged themselves up to the ears, and set off into the hills to make their fortunes. Their first task was to cut, dry and burn the hitherto untouched forest that stood in their way; and though it was backbreaking work, it was certainly achieved more quickly than the long job of converting the blackened, stump-strewn remains into fertile pasture. Inevitably, the boom came to an abrupt end in 1921, largely because the commandeer system was stopped, and many of the new farmers were forced off their land when they had only just got started.

The years between the wars saw a pause in the headlong destruction of the native forest. The steady growth in area of sown grasses declined as the limits of the lowland soils were reached and the hill farms resisted all attempts to break them in. The purely extractive industries (the production of kauri gum and milled timber, the mining of gold and coal) had all passed their best times. No serious attempt at reafforestation was made until the 1920s, and then it was with exotic pines only. No other usable sources of metals were discovered, despite the tireless explorations of surveyors all over the country – many of whom, like Harper and Douglas, also wrote eloquently of the natural life they saw in their travels – and large-scale industrialization was prevented by the lack of any large supply of iron. Even in the mid-twentieth century, more smoke could still be seen belching from the bush than from factory chimneys. The soil on the hills was once stitched down firmly with

In the later European period the pace and extent of land clearance accelerated again. This is the start of a 30,000-acre bush fire at Pukatora Station, 20 km northwest of Tokomaru Bay, in the early 1900s. Alexander Turnbull Library.

Underbrushing and burning the bush was no easy task (try counting the number of cut stems in the picture) but establishing a pasture afterwards was worse, especially as no one understood why the soil that used to support luxuriant bush could scarcely be coaxed to grow grass for more than a few years. Many of the soldier settlers hardly got past this crucial stage when their money ran out. The bush was already destroyed, and when the men left, the impoverished soil reverted to scrub and fern. Alexander Turnbull Library.

*The optimism and willpower of men
who could contemplate the task of
turning this into a prosperous farm,
with nothing but hand tools, muscle
power and fire, makes the modern city
dweller feel limp. Yet they cheerfully
built neat, strong houses in the ruins of
the forest, and most eventually
succeeded in taming the land. This
homestead was photographed in 1890.*
Alexander Turnbull Library.

*In Taranaki in the late nineteenth century cows grazed among the blackened skeletons
of the forest trees, and dairy factories were built to collect the milk, delivered daily by
horse and cart. By 1895 New Zealand had three and a half million hectares of
pasture sown with European grasses, although the problems of maintaining its fertility
were already serious.* Alexander Turnbull Library.

interlocking tree roots; but when it was exposed by fire, and shaken loose by earthquakes, a series of heavy rainstorms were enough to send it 'down to the sea in slips', taking with it most of the country's store of natural fertility. Around Gisborne the rate of erosion from 1932-1950 was five to ten times greater than at any previous time back to 150 AD.[27] It was true that the huge South Island pastoral estates could still make a good living from wool and meat, and in the North, the introduction of the electric milking machine, herd testing, and the early fruits of soil science research in the 1920s greatly increased the productivity of the established dairy farms. But the hills, having repulsed the first onslaught, had earned a temporary respite.

After the Second World War, a new generation of soldier-settlers arrived, equipped with far more formidable weapons, and determined to beat down the last of nature's resistance. Cheap war-surplus aircraft, and experienced pilots to fly them, were available to tackle the hitherto impossible problem of spreading chemical fertilizer and poisoned rabbit-bait over huge areas of rugged hill country. From 1946 the numbers of livestock increased at a truly astonishing rate every year for 30 years,[28] as the battered scrub-covered hills were slowly transformed into the crumpled bright green blanket so much admired by modern tourists. The forests retreated to their last stands, often on the steepest hills where, at last, their value for soil conservation was appreciated. The lowland forests were steadily reduced to a scatter of dark green islands in an alien sea. Most timber production since the Second World War has been based on controlled harvesting of man-made pine plantations, but the old exploitative attitude is even nowadays alive and well in the few remaining companies still holding Forest Service contracts to mill native timber.

An effective answer to the rabbit problem was found after the Second World War. Rabbits were 'decommercialized', and the wages of pest destruction officers were no longer related to the number of rabbits they killed, so there was no incentive to provide for next season's income. Rabbits were killed indirectly and en masse by poison spread from the air. This machine chopped up carrots into small cubes, coated them with 1080 poison (foreground) and fed them into a hopper ready to load into the plane. After 1948 this technique, allied with night-shooting by spotlight from landrovers, finally began to make a real impression on the numbers of rabbits, which 50 years of men and carnivores, working as individual predators killing individual rabbits, had failed to do. W. E. Howard, DSIR Ecology Division.

Predators of the later European period and their impact on the native fauna

The present fauna of predators is rather different from that of previous times. The kiore and the wild dogs are gone, and comprehensive legislation introduced by stages since the late nineteenth century (protecting huia in 1892, kiwis and kakapos in 1896, and all species by 1953) prevents further hunting or collecting by man. The Norway rat now lives mainly in towns, farms, rubbish dumps, streams and swamps. Feral cats still live in the forest, most commonly within reach of civilization, where their numbers are most easily reinforced (even if only temporarily) by strays or by soft-hearted owners dumping unwanted pets. The ship rat, already established in the North Island, spread all over the South Island after the 1890s. The last of all to arrive, the hedgehog, is now very common, except in Westland; despite much lighter traffic, in the 1950s about 50 times as many hedgehogs were killed per kilometre on New Zealand roads than in Britain.[29] They eat mainly worms, slugs, snails and insects,[30] and are too clumsy to be active predators of birds, but they are often accused of crunching up the eggs and small chicks of ground-nesting birds. The mustelids are easily the most significant of the newcomers, and their arrival in force stacked the odds against what remained of the vulnerable native fauna higher than they had ever been.

Just as the violin, cello and double bass form a graded set of musical instruments of the same shape but different sizes, so the three mustelids – the weasel, stoat and ferret– form a size-graded set of slim, active predators. They all have long bodies and short legs, bright black eyes, sharp noses, and small ears. Their habits have earned them a bad press for a hundred years; even in children's books they are always cast as

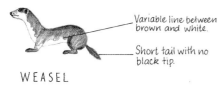

WEASEL

Variable line between brown and white.

Short tail with no black tip.

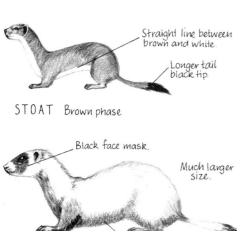

STOAT Brown phase

Straight line between brown and white.

Longer tail black tip.

Black face mask.

Much larger size.

FERRET

Stockier build.

The three members of the carnivore family Mustelidae that came to New Zealand were the weasel, the stoat and the ferret. This drawing points out the differences between the three. When one can get a close look, identification is easy, but in the wild, one usually sees only a low brown streak dashing across a road. The mustelid family has about 70 species, including the otters, badgers and martens of the northern hemisphere; despite centuries of persecution and exploitation for fur, none is endangered except the blackfooted ferret of the American prairies. Cynthia Cass.

villains. Toad of Toad Hall and Little Grey Rabbit had a lot of trouble with them. The reality, however, is not so simple.

Ferrets, the largest animals of the set, are locally common in open country such as farmland, small patches of bush, open riverbeds, tussock grasslands, dunes, swamps, lagoons and coastal areas. In a Wildlife Management Reserve at Pukepuke Lagoon, on the west coast of the North Island, scats (droppings) were collected from live-trapped ferrets during the early 1970s. Remains of birds and eggs were found in 34 per cent of 203 scats, mostly in spring and summer.[31] A general survey of the foods of ferrets collected from all over New Zealand in the early 1960s found birds in less than 30 per cent of guts at all seasons. Ferrets are in breeding condition between September and March, and produce five to ten young at a time after a 42-day gestation, usually in October or November. When food is very abundant, adult females may have another litter in January or February. No doubt many of the ferrets first released onto the barren Otago pastures teeming with rabbits in the last century could produce two litters a season for years. The rate of natural increase of ferrets in those days must have been remarkable.

Nowadays, numbers of ferrets may increase when the local rabbits get the upper hand, and in places and at times when ferrets are common, they could affect the nests and young of waterfowl, seabirds, waders (e.g. the endangered black stilt) and introduced gamebirds. But the population of ferrets is much less now than it was before the rabbits were brought under control in the 1950s.[32]

The stoat is common throughout New Zealand, especially in the bush, and it has also got onto some of the nearest of the inshore islands, including Resolution and Maud. In the mid-1970s, a survey of the biology of stoats in all the National Parks of New Zealand collected 1600 animals in four years.[33] Their guts contained the remains of 2643 meals, mostly taken from birds (in 43 per cent of the guts), mice (19 per cent), rabbits or hares (which are indistinguishable by the time they have been chewed up by a stoat) (18 per cent), possums (10 per cent) and rats (6 per cent). Almost half of them had also eaten one or more insects, mostly the large flightless wetas or native crickets. Stoats varied their diet according to the season of the year and what kind of forest they lived in, and male stoats tended to eat larger prey than females.

Young female stoats become sexually mature at only a few weeks old, and are fertilized by an adult male before they leave the nest – presumably the same one who also serves their mother at around the same time. The release of the ova is stimulated by mating, and the fertilized eggs develop normally for the first two weeks; but then they stop, and pass into a kind of suspended animation known as 'delayed implantation'. Thus they remain, for the next nine or ten months, while the young females grow to adult size at six months of age. Their brothers reach both physical and sexual maturity at ten to eleven months. In the following spring, as the days get longer, the sleeping blastocysts awake, complete their development in four weeks, and the litter (averaging six to ten young, fewer in bad years) is born in late September or October. Most stoats live for only about a year, though individual adults, once established, can live for five or six years.

Stoat populations normally fluctuate through the year, but in most years they are not abundant. Pure meat-eaters living at the top of the food chain are always rare.[c]

But during a mast year in the beech forests, when the forest floor is littered with beech seeds, the feral housemice living in the wild continue to breed and to survive well all winter, so that by spring, when the stoat mothers produce their young, there is plenty of food for them. A few months later, in December and January, an extra large crop of young stoats bursts upon the forest,[34] which can be bad news for the birds still rearing their chicks, and for the newly independent young birds still learning to fend for themselves. Fortunately, these sudden irruptions of stoats last for only a short time: most of the young stoats die within a few months, and by the next spring, things are back to normal again. Unfortunately, we still do not know precisely what effects these events have on forest birds, although we suspect that, for rodents, they can be considerable (p.127).

The smallest of the three, and the rarest in New Zealand, is the weasel. At least as many weasels as stoats were introduced, and their capacity for rapid population increase is greater than that of the stoat; but nowadays they are much scarcer, certainly in all the National Parks. This is largely because their main natural prey, the field-vole, is absent here, and feral house mice are usually not nearly so common in our bush as bankvoles and woodmice are in the woodlands of England.[35] So weasels here tend to be rather erratic in distribution, appearing in some numbers one year and disappearing the next. The average lifespan of a weasel in Britain is less than a year,[36] so most of them do not live long enough to see their first spring breeding season. During the temporary population peaks of field-voles that recur every few years in the weasel's homelands, young females can breed in the summer they were

The weasel is perfectly adapted in every way as a predator of mice and especially of voles. It can follow them down their runways and burrows, and takes over their nests to sleep in, usually lining the inside with the fur of the late occupant. Over most of the weasel's enormous range in the northern hemisphere the winters are long and severe, so the voles are protected from most other predators by the blanket of snow – but not from the weasel. The absence of voles in New Zealand makes it a less attractive place for weasels than for stoats. Cynthia Cass.

born, at about three or four months old, and (since gestation is only 34 to 37 days) their mothers can produce a second litter. In New Zealand, the first litters, of probably four to six young each, are probably born between late September and early November. We do not know if they ever have second litters here, or whether the young born when mice are abundant just survive better than usual.

Weasels eat mainly small prey: 30 dead ones collected during the middle 1970s, mainly from Mount Cook, Tongariro and Craigieburn, had eaten mainly birds (30 per cent) and mice (37 per cent), as well as geckos and insects.[37] However, weasels are too scarce to do any significant damage to birds in New Zealand, at least not in mainland National Parks. Where they are common, as in England, they have caused much annoyance by raiding nest boxes put up for the birds being studied by ornithologists; and in years when rodents are scarce, this predation can affect the population density of box-nesting titmice.[38] But this does not prove that weasels affect the population density of wild birds that do not nest in boxes, and their chances of doing so in New Zealand would be even less. Nevertheless, weasels could cause havoc if they got onto an offshore island with a dense population of small, ground-nesting birds or rare native invertebrates; although this possibility is extremely remote, it exists and should not be forgotten.[d]

Impact on the native fauna

We cannot say how many of the species which seem to have been affected most severely in the later European period had already been distressed by previous environmental changes; no doubt most were. But there are enough reliable, dated observations to show that some species that are now rare, endangered or extinct were still relatively common in the 1870s, and showed serious decline only later.

The list totals 13 species or subspecies (Table 2). Four are certainly or probably totally extinct, all of them confined to the South Island: South Island bush wren, South Island thrush, South Island laughing owl and South Island kokako. Two more, the Eastern or Buff weka, and the South Island saddleback, are now extinct on the mainland but survive on offshore islands. Seven more declined permanently in distribution and numbers after the late 1870s: little spotted kiwi in the South Island, black stilt, brown teal and blue duck, red-crowned and orange-fronted parakeets, and the North Island kokako. The South Island robin and bellbird also declined, but later recovered.

It seems certain that the most extensive change in forest bird populations during this period was not in the North Island, as in the early days of European colonization, but in the South, and especially in Westland and Fiordland. The extent and speed of the changes are clear from the writings of the pioneer naturalists in the south and west.

Rabbits have gone from Westland nowadays, as have ferrets, but in the flush of the first population explosion, the press of numbers forced both to spread out in all directions, even into what afterwards proved to be quite unsuitable habitats. In the short period before they disappeared, they did considerable damage. On the grassy flats along the valley of the Landsborough River, on the thickly forested west side of the Great Divide, rabbits were at plague numbers by 1895; as the explorer A. P. Harper wrote:

The small flat on which we were camping . . . had literally thousands of rabbits, the ground being as bare as a barrack yard. When we reached this open space and came out of the trees on to the grass, it seemed as if the whole surface of the ground turned a somersault in sections – in such countless numbers were the rabbits diving into their burrows. The ground looked honeycombed . . . they must have come from the eastern side of the Range via some low pass.[40]

Naturally, the mustelids followed them, and so discovered the last remaining rich fauna of ground birds left anywhere in the inhabited world. The result of this encounter was easily predictable. The Chief Surveyor for Westland, George Mueller, wrote in his report for 1889/90:

During the past summer several weasels[e] and ferrets were caught and killed at Okuru and Waiatoto settlements [on the Westland coast] . . . nobody introduced them into Westland, and hence they must be progeny of those imported by the Government, and must have found their way across the Dividing Range from either Otago or Canterbury or both. But in the absence of any signs of rabbits about the coast settlements it is difficult to understand what brought these animals over. This mystery was effectively cleared up on my exploration trip. . . . The ferrets and weasels came up to the Dividing Range with the rabbits, but as soon as they discovered our ground birds – our kakapos, kiwis and woodhens, blue duck and suchlike – they followed up the more palatable game[f] . . . henceforth the rabbits on our side of the Dividing Range will be left undisturbed and allowed to spread as they please . . . but as regards the ferrets and weasels, etc., they will thrive and will continue to thrive until the extermination of our ground birds, which is now begun, is fully accomplished.[42]

And in 1890, the *Otago Witness* reported that 'in the Hollyford Valley [was found] a ferret warren, and the weka, kiwi and kakapo were almost exterminated. In the Makaroa Valley[g] these used to be plentiful, but since the advent of the stoats and weasels they are very rare, and rabbitting tallies have not depreciated.'[43]

Charles Douglas (left) and A. P. Harper, with Douglas's dog Betsy. These two famous early explorers ranged the West Coast in the second half of last century, and their writings give us some glimpse of how different the wildlife of the Coast was then compared with now. April 1894, Cook River. Canterbury Museum.

Charles Douglas was an indefatigable explorer of the rugged mountains of Westland, and a profound admirer of the native wildlife he found there. He often spent months at a time surveying in remote back country seldom if ever previously visited by man, and he was much struck by the tameness of the birds there, just as Cook had been at Dusky Sound, and as Darwin had been in the Galapagos. During his nearly 40 years in the bush he watched, with helpless sorrow, the unequal contest between the unsuspecting native birds and the various alien killers brought in by the settlers. In *Birds of South Westland*, Douglas described the birds as he first knew them in the 1860s:

> Sometimes a dozen thrushes will be round a tent eating off your plate, or even out of your hand, and if the camp is shifted, they will follow it to the next place if the distance is not too far. . . . They take food out of a man's hand with an indifference to his presence, as if he was merely a stuffed effigy of some sort, or a kind of walking tree, that grew crumbs and bits of butter.[44]

But by 1891, Buller was reporting that 'The South Island thrush . . . is still comparatively plentiful in some parts of the West Coast, but its numbers have been grievously diminished by the diggers' dogs, by wild cats, stoats and weasels.'[45]

On the kakapo, Douglas wrote:

> Although so formidable looking, the Kakapo appears to have little idea as to how to defend itself against dogs, ferrets or men. If a dog puts its nose, or you put your finger into the claws or beak, you will both know it, and be more carefull in future. If the bird only knew its powers, it wouldn't fall such an easy prey [to] stoats and ferrets. One grasp of his powerfull

The native thrush or piopio, *a ground-feeding bird that has lived in New Zealand since well before the Pleistocene. Like many of the birds of undisturbed islands, piopio appeared to be absurdly tame to the first human explorers to encounter them: this drawing illustrates Douglas's comments about them, quoted above.* Cynthia Cass.

claws would crush either of those animals, but he has no idea of attack or defence . . . in a good kakapo country before the advent of the ferret and the stoat . . . the birds used to be in dozens round the camp, screeching and yelling like a lot of demons, and at times it was impossible to sleep for the noise. The dog had to be tied up or matters would have been worse. It would have been killing and fetching all night long, but alass, this is a thing of the past. . . .[46]

By the 1890s, Douglas's memories of the abundant birdlife he used to know were becoming almost sentimental:

Years ago the Karangaroa and other rivers in Southern Westland were celebrate for their ground birds . . . The Weka prowled round the Tent, anexing anything portable and the Kiwi made the night hedious with its piercing shriek. The Blue Duck crossed over to whistle a welcome. The Caw Caw swore and the Kea skirled, Piegeons, Tuis, Saddle backs and Thrushes hopped about unmolested. The chorus of the Bell bird was heard in the dawning and all were tame and inquisitive, but now all this is altered. The Digger with his Dogs, Cats, Rats, Ferrets and Guns has nearly exterminated the Birds in the lower reaches of the southern rivers. The cry of the Kiwi is never heard and a Weka is a rarity. The Blue Duck once so green [naive], is as carefull of himself as the Grey and the Robins are extinct.[47]

But on a visit to the Copland Valley in 1892, he had a pleasant surprise:

. . . the Flats of the Copeland put a fellow in mind of old days; it was full of birds all tame and inquisitive as of old . . . the Robins ate out of your hand and the Bell Bird sung its chorus in a style only now to be heard below Jacksons Bay and the Blue Ducks were tame as of yore.[47]

Unfortunately, this happy respite did not last long. In 1895 Harper visited the same area. He wrote:

It is hard to believe that birds could disappear so quickly as they have in this valley. Compare Douglas's picture of peace and plenty [quoted above] with mine three years later. I should say that never, with the exception of Cook River and the Twain Valley, have I seen such a dearth of birds – of kiwis we neither saw nor heard a trace, of wekas we caught two and saw one. Dick says he heard one robin, which is more than I did – bell-birds were either non-existent or silent – of blue-ducks we saw one pair, so wild that we could not get near them. Whereas Douglas caught and shot some thirty wekas and between twenty and thirty ducks for food on the river generally, and left hundreds, we only got three kakas, two pigeons, and two wekas; and instead of, like Douglas, finding too much to eat, and having to leave stores behind rather than bring them out, we took more with us than he did, and yet were on short rations for two days. Douglas was the first man in this valley, and between his visit and ours (except Fitzgerald, who did not attempt to catch any) *no man had been into these solitudes. The decrease must be entirely due to cats, and to a greater extent to weasels'* [his italics].[48]

The dearth of birds in the Cook River had caused Harper some anxious moments in the previous year. The explorer's traditional practice of supplementing food supplies with birds, so as to reduce the load to be carried, had been reliable since the 1860s; but after about 1890 the birds disappeared so completely, and so entirely without warning even in virgin territory, that some expeditions were caught unprepared. Harper's protestations, that his troubles in 1894 were not the result of bad planning, emphasize how sudden the change must have been:

It may, perhaps, be thought that we only had ourselves to blame for short rations and starvation on this trip, but I think it was our misfortune, not our fault. In the first place, the valley was unexplored, and we had every right to look forward to as many birds as we had need of for food; and, as we always rely greatly on these, we only took enough food to last us for the trip, *with help of birds.* Again, we did not anticipate more than ten days' work at the most, so we took flour, rice, oatmeal, tea, cocoa, sugar, a little meat, treacle, suet (for cooking dough-boys), and a tin or two of sardines in sufficient quantity, *plus* birds, to last us for that period. Had we found birds, as we reasonably anticipated, the provisions we took would, with care, have lasted more than two weeks, and even if they were exhausted, we could have lived well with the help of the pea-rifle. Luck was against us in every respect; for the first three or four days we had meat, and went on eating as if there was no need to economize. By that time we had gone some way up the river, and the bad weather not only prevented a retreat, but delayed our advance. . . . It is no joke to be compelled to divide six good meals consisting of flour and rice into rations to extend over ten days, and at the same time do a considerable amount of heavy work.'[49]

Further north, the same dismal reports came from the remote forests inland from Nelson. In 1895 Buller quoted one of his correspondents, who wrote to him from the Pelorus River:

. . . I am camped a long way up the creek, at a place where I used to collect birds some years ago. In those days I found this a good hunting ground; a great number of species could then be obtained in this locality; but now all this is changed. I seldom see or hear any birds worth collecting. The stoats and weasels have done their work . . . I do not now see or hear any saddlebacks, or pigeons, or wrens, all of which were plentiful enough in this place a few years ago. The Blue Duck used to be fairly abundant in the creek, and they are now nearly extinct. I attribute this to the stoats, which are very numerous about here. Collecting specimens of natural history in this part of the country is a thing of the past, for the stoats and weasels have swept away everything.[50]

The lower reaches of the Haast River in South Westland. The Landsborough River is one of its upper tributaries; the Copland and the Cook Rivers, both further north, flow through equally rugged mountains. With the inadequate camping and climbing equipment of their day, the early explorers endured many hardships amid this marvellous scenery.
Author.

Much the same conclusions have been reached by writers in modern times. In *Dusky Bay*, the Begg brothers made a careful comparison between the birdlife of Dusky Sound as recorded by Forster in 1773 and by Henry in 1895, and as they observed it in 1963. They concluded that:

> most of the bush birds are still to be found at Dusky Sound despite the depredations of the hawk, the stoat and the rat. It is the ground birds which have suffered. The kokako, the piopio, the robin, the saddleback and the kakapo, all ground feeders, have been reduced to the point of extinction. The weka and the kiwi, though still surviving, are comparatively few in number. We saw a stoat at Cascade Cove within five minutes of our arrival. It is a reasonable deduction that these introduced predators, together with the rats, are responsible for the destruction of many of our unique ground birds ... Richard Henry reported a weasel on Resolution Island in 1900, and other visitors have repeatedly noted their presence in the Sound. Mr Ken Sutherland trapped twelve stoats in Cascade Cove during short visits in 1957 and 1960. As rabbits have never penetrated as far as Dusky Sound, the main diet of these ruthless hunters must be birds.[51]

In *Port Preservation*, the Beggs undertook the same exercise for Chalky Inlet, comparing the birds recorded by Andreas Reischek in 1887 with those they found in 1969. The conclusion was the same, and expressed slightly more strongly:

> ... during the eighty-two years which elapsed between Reischek's observations and ours, the number of native birds and the distribution of different species has changed considerably. The birds which lived and fed on the ground – kiwis, kakapos, wekas, saddlebacks, piopios, kokakos and robins – have now disappeared, confirming Reischek's prediction that the ground-inhabiting birds 'had little chance of preserving their species'. [*see p. 70*]. Voracious mustelids have completed the destruction begun by the rats and the miners' dogs. We saw both a rat and a stoat hunting on the edge of the beach within a few hours of our arrival at Preservation Inlet.[52]

In *Hidden Water*, Philip Houghton quotes the list of birds common in Martin's Bay in the 1870s, compiled by Jane Robertson, who lived there from 1870 to 1887. Among her father's activities was the collecting of bird specimens for the Otago Museum, a task with which the whole family assisted. Over the seven or eight years up to 1881, they prepared hundreds of skins and skeletons, and they must have known the local birds very well. Among the species listed by Robertson, which Houghton seldom or never saw in 16 years (1958-74) of regular visits to the same area at all seasons, are kokako, piopio, parakeets, saddleback, kakapo, weka and kiwi. Houghton considers various reasons for their disappearance, concluding:

> It is only with the appearance of the stoat that the birds seem to have declined in numbers ... stoats are thick everywhere here. One might fairly think that the high rainfall and wide rivers would hamper their multiplication and spread. But they thrive and swim ... I pace a silent bush and curse the memory [of the man who thought up the idea of introducing stoats to this land], for deer, rabbits and opposums have not robbed us of one fraction of our heritage that the stoat has.[53]

It is clear that the contemporary observers who witnessed the effects of mustelids – especially stoats – on the native fauna of Fiordland and Westland around the turn of the century were in no doubt that these predators were among the very worst ever to have been brought to New Zealand. Then and now, people have tended to assume that such drastic effects were also going on elsewhere, and that therefore stoats must have made a very significant contribution to the historic extinctions over the whole country. This goes far towards explaining the hostile attitude of writers and naturalists towards stoats nowadays, exemplified by the Beggs and Philip Houghton, and by the correspondents quoted in Chapter 5. Houghton's assessment of the role of the stoat is perfectly understandable – but is it, in fact, true?

The contribution of stoats to the *total* history of decline of the New Zealand avifauna can be estimated quite objectively, because they came late and are confined to the main and closest inshore islands. Species that were gone from the main islands before they arrived, or which have gone from islands they never reached, could not have been affected by them.

Altogether, *at least* 153 distinct local populations of birds of the New Zealand region are known to have become extinct, or greatly reduced, in the last thousand years. If they are listed according to whether or not they ever came into contact with stoats,[54] we see that, in the broad view, the part played by stoats in the *total* history of extinctions in New Zealand appears almost insignificant. Of 135 separate populations, now certainly or probably extinct, only five (four per cent) could have been reduced, or at least finished off, by stoats (Table 4), and even these could in fact have been affected more by ship rats than by stoats. Of 18 separate populations whose range and numbers have been greatly reduced, and are now regarded as rare or endangered, 11 (61 per cent) could have been affected by stoats and/or rats. There is not a single known extinction or diminuition in New Zealand that can be regarded as definitely and solely due to any of the mustelids, despite all that has been written about them as agents of terrible destruction. An authoritative survey in 1973 concluded, 'It is actually difficult to attribute the decline of any native bird directly to mustelids'.[55] Overseas the story is the same: only one per cent of 163 extinctions, recorded from islands all over the world since 1600, have been attributed to mustelids, compared with 26 per cent attributed to cats and 54 per cent to rats (Table 5).

This conclusion is certainly not the same as saying that the stoats brought to New Zealand were not capable of predation heavy enough to cause extinctions. It is largely that they did not have the opportunity. Over the whole of New Zealand, although the total list of extinct and endangered birds is shockingly long, many of the species listed lived on the offshore and out-lying islands, reached by some of the other agents of change, but not by stoats. Barriers of time and geography prevented stoats from even meeting 137 of the 153 distinct local populations of birds that have drastically diminished or disappeared. It is irrational to include stoats in discussions of the causes of the decline of native birds which were never exposed to danger from them.

On the mainland, the historian's evidence that steep declines in numbers of birds soon followed the spread of mustelids into the back country comes mainly from the South Island, and is entirely circumstantial. In fact, if our three weasely friends were in the dock on a charge of murder, the chances are that they would have to be acquitted. Why? Because of a common frustration of historical research – the unfortunate coincidence which can confound and undermine even what appears to be cast-iron evidence. In this case, the coincidence is that the two predators that are potentially the most destructive to birds that we know of, the ship rat and the stoat, probably moved into the South Island at about the same time, so we have no way now to distinguish between their effects, nor between the new changes they made and the accumulated changes wrought by the Norway rats and cats already present.

Ship rats are deadly to birds, but nocturnal and inconspicuous, and seldom specifically mentioned by the naturalists of the 1890s. There is some case for the argument that ship rats alone could have destroyed the birdlife of Westland and Fiordland quite unaided, as they could have done in the North Island earlier, and certainly have done on various other much smaller islands (Chapter 6). Because the attention of contemporary observers was naturally directed towards the much more obviously predatory mustelids, the arrival of merely another kind of rat was less often remarked on, and its part in the slaughter could have been seriously underestimated. On the other hand, one could argue that effects as drastic as on, say, Big South Cape Island (p.71) were possible only because those rats, free of predators, were able to reach enormous numbers.

We do not know for sure what population densities ship rats reached in the 1890s and 1900s. In New Zealand forests today, they are certainly widespread, especially in podocarp/broad-leaved forests, and they do increase in numbers after the forest trees have produced a good seed crop (recent estimates suggest an increase from a normal one to three per hectare to at least ten per hectare).[56] But even after a massive seedfall, they never reach the 'prodigious' and 'immense' numbers commonly mentioned by naturalists writing before the mustelids arrived. Likewise, the periodic massive irruptions of kiore in the South Island stopped after about 1888.[57] Stoats could not have prevented the fat, well-fed kiore breeding over winter and becoming much more numerous by spring, but that very increase would have ensured large numbers of stoats around in summer, which could certainly have curtailed the irruption of kiore and made it much less noticeable to human observers. That is what seems to happen to post-seedfall irruptions of mice in beech forests nowadays (p.127). So it seems quite possible that it was the spread of mustelids – especially stoats into the forests – which has reduced the average level around which rat populations (of all species) vary. Since then, ship rats (now certainly the commonest of the three in forest) have helped to keep up pressure on birds in two ways – by helping to support the population of stoats, as well as by occasionally doing a bit of thieving on their own account. At present it is impossible to decide which effect is the worst.

This complication does not, however, deny the considerable probability that it *was* predators, including stoats, that caused the sudden disappearance of birds from the untouched forests of the south and west in the mid 1890s. But we do have to think only in probabilities: we cannot be sure that the losses were due to the immediate impact of the three mustelids and the ship rats, because we cannot exclude the possibility that they were already inevitable, the fruition of longer-term campaigns waged by cats, dogs and Norway rats. All we can say is that there were seven predators present in the later European period (Table 2), and stoats probably made a substantial contribution to their combined effect. It is circumstantial evidence, but it is persuasive, and Houghton may be right, at least in Fiordland.

In the North Island, on the other hand, the picture is hopelessly confused. Not only had ship rats already had a start of some 30 years on the mustelids, but also, the human population was greater – and had been for hundreds of years – which meant more forest disturbance and hunting by man and his camp followers, especially rats, cats, dogs and pigs. Worst of all, by the time the mustelids arrived, large-scale clearance of the North Island forest by Europeans was well under way. The contribution made by the addition of mustelids to this scene of mayhem is now impossible to sort out, but the chances are that it was much less than is commonly believed. All the other immigrant killers had already done their worst in the North Island long before the mustelids got their turn.

There were some later declines which are sometimes linked with the spread of mustelids – for example, that of the brown teal in the 1920s and 1930s; but the evidence is extremely shaky. Brown teal also disappeared from the Chatham Islands in 1930, where mustelids could have had nothing to do with it; and on the mainland, by 1930 the confusion of equally likely possible causes (e.g. destruction of forest by deer, disease and competition from introduced birds) has become a fullgrown Gordian Knot. Is it possible to work out what effect these other changes were having on the native fauna?

Alternative explanations

In Chapter 2 we saw that the theory of island biogeography can be used to give a *very* rough estimate of the number of bird species that we would expect to disappear simply as a result of deforestation alone, all other things remaining equal. In Polynesian times, the actual number of birds lost from the North Island (29 per cent) hugely exceeded the number expected from the theory, which was less than five per cent (Table 3). The difference probably reflects the disproportionate effects of the introduced predators on a totally naive fauna. By contrast, between the early nineteenth century and the present day, the number of North Island land birds has fallen by 25 per cent, whereas the theory predicts a loss of about 16 per cent. The difference between the observed and the expected loss is still great, but not as great as it was in Polynesian times. Does this mean that the more recently arrived predators were less damaging? No, it does not. It means that the avifauna remaining when the Europeans arrived contained fewer species extremely vulnerable to predation. It may also mean that there are more North Island extinctions to come.

Only 10 full species are counted as having become extinct in the North Island since 1800, including seven that still survive in some numbers, either in the South Island or on offshore islands. Two others, the blue duck and North Island kokako, are rare on the North Island mainland, and like most endangered species everywhere, some individuals are surviving in what is probably sub-optimal habitat; and even in the last few patches of optimum habitat, the full effects of the rapid destruction of forest in European times are probably still to be felt. One of the most basic ideas of biogeographic theory, repeatedly asserted, is that if only a fraction of a previously large area of forest is retained as a reserve, many more species will survive there for the first few years than can remain indefinitely. So there is a delayed reaction to deforestation (which takes longer to complete the larger the area) by which the initial excess of species is gradually reduced by extinctions, a process known to biogeographers by the graphic term 'faunal collapse'.[58] This somewhat pessimistic prospect is a thought-provoking counterpoise to the misguided hope that, since the Polynesians had already disposed of the most vulnerable species, the rate of extinctions in the European period should be lower. The disappearance of species vulnerable to hunting in Polynesian times is unrelated to the disappearance of species vulnerable to habitat destruction now; the losses are additive, not compensatory.

However, there are at least four objections to the very idea of using this sort of calculation to make even the grossest distinction between the effects of deforestation and of other disturbances, such as predators. The first is the basic statistical caution that correlation does not imply causation: area is certainly *related* to the number of species, but that does not necessarily mean it *determines* the number. The confidence limits about the relationship are wide, which means that it is not much use for making specific local predictions.[59] The second is that, in the simple calculations above, we have summed all areas of native forest on the North Island, whereas, even in primeval times, there never was a single block of forest of 10.9 million hectares.

The third objection is that the rule of thumb was itself worked out, quite recently, from observation of birds in contemporary forests which contain the contemporary range of predators, and so the 'rule' already has the effect of predation worked into it. It may be true that, in modern times, if we retain ten per cent of the habitat we may expect to retain 50 per cent of the birds, but that does not tell us exactly what proportion of the original birds would still be present in what is left of the original forests if the predators had never arrived. The relationship between losses of forests and of birds would be different without the predators, and we have no way now of measuring it. The great value of the species-area relationship is in alerting us to the dangers, the probable results of *further* loss of forest – results which may well include, in some unknown way, the actions of predators. We can say that, since nine of the native species still living on the three main islands are already rare or endangered, and man's intrusion into the remaining forests has by no means stopped, we have no guarantee of retaining even what we still have. By refusing to reserve, albeit for valid economic reasons, less than the total amount of remaining forest, we are greatly increasing the chances of further avoidable extinctions, with or without the help of predators.

The fourth objection is that not all forest is of equal value for bird life, and the 6.2 million hectares left is scattered over the whole country, and is not a representative sample of the original 21 million hectares. The remaining forest could not contain the number of bird species, diminished in proportion with the forest area, that the theory suggests, because the formula assumes that all other things are equal. The introduction of predators is not the only reason why all other things are definitely *not* equal. Proportionately far more of the once-vast warm lowland forests, the richest in plant and bird life, have gone or are reduced to scattered remnants. The cooler mountain forests, generally of less value to birds, still exist in large blocks on the major ranges; but the fact that there are still vast areas of high-altitude forest is no comfort to the birds that could only live or winter in the lowland forests which are now gone.[60] Moreover, only 1.6 million hectares (about 25 per cent) of the total remaining forests are protected in National Parks and various other kinds of reserves, and in these the predominance of mountain over lowland forest is even more pronounced.[61] All this is understandable, since what is left of the lowland forests contains not only most birds (including rare and endangered species) but also most merchantable timber, and logging men resent and resist the 'locking up' of such areas in reserves.

On top of these four objections to the use (rather, misuse) of biogeographic theory to predict too precisely the number of extinctions that might be due to deforestation alone, there are at least two reasons, other than the introduced predators, why even the most extensive forests that remain are less useful to native birds than they might be. There are more ways to kill a cat, so the saying goes, than choking it with cream. Likewise, there are more ways to ruin a forest (and its inhabitants) than chopping it down. One of the alternatives is to introduce alien browsing animals.

The New Zealand birds, especially the long-resident, specialized endemic species, had become very closely adapted to the particular characters of the original habitats that their ancestors settled in. The composition of the forests and other natural habitats was as important to them as the fact that they were there. The combination of plant species present in the undisturbed state used to produce a predictable set of resources on which long-established birds had come to rely. For example, there was a variety of fruits and insects which became available at different times of the year in a dependable sequence; there was mountain forest to visit in summer and lowland forest to shelter in in winter. When some of these seasonal supplies or refuges disappeared, there were times of the year when certain birds had nothing to live on and nowhere to go. Then the forest became uninhabitable for those birds all the year round, even though the trees still stood.

Most of the native forests in the main islands that have not been logged look beautiful and natural, but in fact very few of them are still in anything like their original state. To see the difference, you have to visit an island, such as Little Barrier, or the Hen and Chicken group in the Hauraki Gulf, eloquently described and photographed by Gordon Ell.[62] There are several islets in Lake Manapouri, where the luxuriant vegetation on the ground – ankle deep mosses, ferns and native orchids – has obviously never been trampled or eaten, and there are tender palatable plant species that have long gone from the mainland. In Captain Cook's time the forests around Dusky Sound were 'so overrun with supplejacks, that it is scarcely possible to force one's way among them . . . the soil is a deep black mould, evidently composed of decayed vegetables, and so loose that it sinks under you at every step'.[63] Both supplejack and leafmould are still there, but nothing like so thick. The changes in the forest in European times are the work of the introduced herbivores, especially those that browse in forests rather than graze on grass.

New Zealand's vegetation evolved, like its birds, in isolation. All the conifers, 83 per cent of the flowering plants, and 41 per cent of the ferns are endemic species, making up a flora more like the Mesozoic forests of ancient Gondwanaland than any other in the world now living, and unused to being eaten and trampled on to the extent it has been in the last 200 years. (Browsing by moa was not necessarily the same process as browsing by deer, so it is not certain to what extent deer have replaced moa as agents of natural selection for native vegetation.) The potentially damaging effects of deer browsing were pointed out long before deer became numerous, by Rev P. Walsh, in a prophetic paper read before the Auckland Institute in August 1892:

The European forest or deer park . . . has grown up subject to the presence of ruminants of various kinds . . . the floor of the forest is generally covered with a quantity of grasses, fern and brambles which spring up every year, and which amply supply the wants of the animals. But in the New Zealand bush the case is quite opposite to this. The forest has grown up through the course of ages undisturbed by any four-footed enemy whatever. In its virgin state there is no grass, properly speaking, at all, while the undergrowth of ferns, shrubs and seedling plants, once destroyed, can never be restored . . . lamentable ruin has been brought about in a very few years mainly by the cattle of the settler . . . wherever the deer have found a home their ravages are even more rapid and fatal. . . . It is not to be supposed that our bush can forever be wrapped up in cotton wool, so to speak. However we may lament, by far the greatest portion of it is destined to fall under the axe of the timberman and the settler . . . still, an attempt might be made to do something to preserve at least a few limited areas of the forest in its virgin state. . . .[64]

Walsh suggested the simple, effective policy of fencing the reserves which were then being set aside, and setting an armed guard on them. But the policy-makers of the time were not ready for such drastic action, and they dismissed Walsh's plea on the grounds that fencing large reserves was impossible and that destruction of New Zealand's forests was inevitable. It is left to us now, whenever we look at the rampant growth inside an exclosure in the bush, to wish most fervently that they had listened.

Wherever a herd of red deer got properly established in the bush, a predictable sequence of events followed.[65] After a slow start, their numbers and distribution increased more and more rapidly, with the deer in splendid condition and the stags

The Acclimatization Societies enthusiastically sponsored repeated liberations of red deer in New Zealand, which started in 1851. Their aim was to make up for nature's carelessness in omitting to provide the country with game animals. They had no idea that a positive explosion in numbers of deer would be having drastic effects on the forests within 50 years: intensive browsing and ringbarking practically eliminated many understorey shrubs and trees, like this fivefinger, which were important sources of fruit and berries for the native birds. Cynthia Cass.

At peak numbers, deer can remove the entire understorey of a forest, leaving the ground bare and trampled. Food and cover for native birds are completely destroyed. The effect can be graphically illustrated by exclosure experiments. This plot, in the Aorangi Mountains of the southern North Island, was fenced in 1950; in 1962 the thick regeneration inside reached above a man's head. W. E. Howard, DSIR Ecology Division.

growing huge trophy heads of antlers. New Zealand's equable climate provided a long growing season and a year-round evergreen food supply, so the growth of the population was not constrained by the annual bottleneck of the northern hemisphere winter, which sets a limit on the number of animals that, having lived through the season of food shortage, are still around to browse the fresh growth of spring. After about 20 or 30 years, the densities of the herds were immense, but both the animals and the bush were in a very poor state. The stags grew only small or malformed antlers; the does frequently failed to rear their fawns; there was heavy mortality over winter, and general starvation so weakened the animals that they sometimes fell over as they attempted to run away. The native forests developed, to various degrees, signs of severe to chronic overbrowsing, such as cleared-out undergrowth, trampled and uprooted ground cover, lack of regeneration, compacted sub-soils and exposed tree roots. Such a devastated habitat could not support the huge numbers of deer, so the population inevitably fell back to a more sustainable level. In the northern South Island, this sequence of events has been reconstructed as follows:[65]

1861	Initial release near Nelson, slow early spread
1900	Range 5000 km², all at low density; deer in good condition
1905-1930	Increase phase, expanding to range of over 30,000 km², 39 per cent of which was occupied at high or very high density; deer in deteriorating condition
1930-1945	Peak populations, covering about 35,000 km², of which 47 per cent was at high or very high density; deer in very poor condition

1945-1960   Decline phase, during which the total population fell back to about what it had been around 1910, covering a very large area but intermittently or at low density: condition improving.

The other most important browsing mammal to be introduced was the Australian brush-tailed possum, which spread out in overlapping waves from many liberation points in both islands, just as did the deer, and over roughly the same period. Possums were freely liberated and protected until 1921, and for the next 25 years they were harvested under a restricted licensing system. By 1942, however, there was growing uneasiness about the increasing number of possums, and about damage to forest trees in Westland.[66] After 1946 the possum was recognized as a potentially dangerous pest, and the legislation was changed to permit trapping and poisoning for control rather than for fur. Millions of dollars are now spent every year in possum control work.

The effects of these events on the mainland native birds are not known in detail, but easy to guess. Red deer and possums are a devastating combination when in high numbers, each browsing on species out of reach of the other. On the ground, deer ring-bark small trees with palatable bark, remove much of the undergrowth, disturb the leaf litter and prevent the regeneration of palatable shrubs and ground ferns.[67] Up in the canopy, possums browse leaves of rata, kamahi and fuschia, often

The Australian brush-tailed possum was introduced for its fur, and is now abundant throughout most of the country. But like deer, possums have strong preferences for certain trees and shrubs. One of their favourites is the rata, whose flowers provide abundant nectar for honey-eating birds. Possums at high density can kill a mature rata in only two to four years.[68] Cynthia Cass.

concentrating on individual trees, biting off every new shoot until the tree is exhausted. [68] After a few years, the plants most favoured by deer and possums disappear entirely, and they are replaced by different, browse-resistant species. Unfortunately, many of the most tender plants are the very ones most important to birds. Rata and fuschia flowers provide nectar for the honeyeaters (bellbirds, tuis and stitchbirds);[69] understorey shrubs and leaflitter provide the berries, and a rich habitat for the insects, fed upon by other birds. Selective removal of such species by deer and possums leaves a gap in the annual food budget of these birds which may be hard to fill.

Further up the mountains, the subalpine scrub is often badly damaged by deer sheltering there during the day, ready to move out onto the open tops under cover of darkness. In Mount Aspiring National Park this zone is now dominated by exotic birds, particularly the chaffinch, redpoll and hedge sparrow. Peter Child[70] lists four species of birds native to that park as likely to be seriously or considerably affected by the destruction of the scrub and understorey layer. Two are now probably extinct (South Island kokako and bushwren) and the other two (fantail and tomtit) are now less common than the introduced chaffinch. The four native species least affected by the activities of the deer are those that feed in the canopy or on the trunks of trees: the grey warbler, rifleman, brown creeper and yellowhead. The grey warbler and rifleman are now the commonest native species in the park.

Degradation of the habitat is therefore the first of two reasons why many native birds species have disappeared even from the considerable tracts of forest that remain. This implies that rigorous control of browsing mammals should be one of the foremost policies for the conservation of our remaining native birds, at least in the National Parks. Unfortunately the question of control of deer is a complex and highly emotional one, for reasons explained with disarming logic and entertaining candour by Graeme Caughley in *The Deer Wars*. The complexity of the problem is less to do with the deer than with how people view them and the forests in which they live. New research suggests that deer and possums may not be quite as villanous as they are painted; and anyway, in the long run the controversy will resolve itself, probably on nature's terms. Although deer do modify the vegetation, the process does not go on indefinitely. An equilibrium is achieved after about 40 years, at which the vegetation (and the birds remaining in it) reach some sort of accommodation with the deer – a different, but equally viable ecosystem will evolve. The fears of Walsh and other older naturalists, that the forest would be modified to destruction, are unfounded.[71] Nevertheless, from the birds' point of view, the new situation is unlikely to be an improvement on the old one.

The effects of introduced browsing mammals are most dramatically illustrated on some of the outlying islands, where the worst culprits are usually goats, not deer. For example, goats were introduced on to Macauley Island sometime before 1836. By 1887 they had transformed the indigenous scrubland to close-cropped, eroding grassland. In 1966 the goats were exterminated, but by that time virtually no woody vegetation or forest birds remained. Now the native shrubs and trees are regenerating, including some species which had long vanished, and birdlife has already shown a marked recovery.[72] Macauley Island is an extreme example, but there are places on the mainland, such as the rocky mountain ridges of the Ruahine

Macauley Island, 300 hectares, is one of the subtropical Kermadec Group of islands, which lie 1000 kilometres northeast of New Zealand. The goats liberated here to provide food for seamen took only 50 years to destroy the native scrubland. When the N.Z. Wildlife Service decided to rehabilitate the island, over 3000 goats were shot. J. Clark, DSIR Ecology Division.

ranges, where goats are common enough to make an important contribution to the modification of the forests. In the Mapara and Hunua forests of the central North Island, browsing by goats may be seriously affecting the survival of the last few kokako. The total number of goats shot in control operations since 1931 exceeded one million by 1970.[73]

Some writers have suggested a second reason why the native birds can be disadvantaged even in the relatively vast forests that remain: they now have to compete with the introduced birds for living space. Of the roughly 130 species of foreign birds introduced or escaped from captivity since 1840, many of them already adapted to a forest habitat in their homelands, 36 have become established. Of the very many other species that visit our shores annually after unassisted passages, usually from Australia, ten (mainly water birds) have become established in the same period (Table 6). The relationship between the extinctions and colonizations of the last 140 years has been much debated. Did competition from the exotic species contribute to the extinctions of the natives? Or was the establishment of the exotic species permitted by the extinction, from other causes, of the natives?

Charles Darwin himself wrote, in *The Origin of Species*, that 'from the extraordinary manner in which European introductions have recently spread over New Zealand, and have seized on places which must have been previously occupied, we may believe, if all the animals and plants of Great Britain were set free in New Zealand, that in the course of time a multitude of British forms would become thoroughly naturalized there, and would exterminate many of the natives.' But G. M. Thomson, after quoting that passage, goes on to say 'I think it will be shown that where man does not interfere with the vegetation, the indigenous species can hold their own against the imported forms. It is human intervention – either direct or indirect – which completely alters the conditions.'[74] It says a lot for Thomson's independence of mind that he could trust his own observations rather than Darwin's immense authority. Not only that, but it seems that he was right. Careful comparison between the mainland and some of the offshore islands of New Zealand suggests that the establishment of exotic species is one of the consequences of man's disturbances of the native forests, and does not, in itself, cause extinctions among native species.

In the mainland (North and South Island) forests, six of the 16 native passerine species (perching birds) present 200 years ago are now either absent (extinct, or confined to offshore islands) or rare and local.[75] Some five introduced European species, plus the silvereye (which introduced itself from Australia in the 1850s), are common even within large tracts of closed native forest. Repeated censuses by J. M. Diamond in 28 mainland forests showed that native passerines accounted for only 64 per cent of all bird individuals seen, and native non-passerines (e.g., pigeon, kaka and parakeets), another 5 per cent. Five European passerines (song thrush, blackbird, dunnock, redpoll and chaffinch) accounted for 18 per cent, and silvereyes alone 13 per cent. In open habitats, pine plantations and town gardens, these strangers are far more abundant than the native species.

By contrast, on Little Barrier Island (3076 hectares), there are no browsing mammals, European rats or mustelids, only kiore and (until recently) cats. The forest has a thicker leaflitter, a denser and more varied understorey, and a greater abundance of native food plants than mainland forests. Little Barrier supports every native resident forest bird species that is widespread on the North Island mainland, some (e.g. parakeets, whiteheads, bellbirds and tui) in much greater numbers than there; plus the sole surviving population of stitchbirds (p.184). Introduced birds are practically absent from the forest, although nine species (including silvereyes and four of the five European species common in mainland forests) breed in the 20-hectare clearing round the resident ranger's house. Evidently, introduced species that are able to penetrate mainland forest, and to reach Little Barrier Island, are excluded from the forest there.

Two smaller islands nearby (Lady Alice, 138 hectares, and Mauitaha, 20 hectares) support only forest that is regenerating after recent burning, and their birds are a mixture of native and introduced, intermediate between those of the mainland and Little Barrier. A third nearby island, Cuvier, once supported cats, wild goats and feral stock, and at least five non-native breeding birds (four European, plus silvereyes). After eradication of cats and goats, and fencing of stock by 1964, a dense understorey has regenerated, and the five non-native birds have almost or entirely disappeared. On Big South Cape Island, introduced species were rare before the 1964 rat invasion removed five species of native birds previously resident there in considerable numbers (p.71): since then, the introduced species have become more common.

These figures suggest that native birds secure in undisturbed native forests are able to repel the invaders from overseas. Only when the original forest structure is changed by logging or browsing, or when the native species are decimated by these changes in forest structure, or by predators, can exotic birds become established. It seems that both the forest changes and the predators are important. Exotic species invade the modified forest on Lady Alice and Mauitaha Islands even though they have fewer predators than Little Barrier, whereas on the mainland, introduced predators are an important factor in the reduction of certain native birds, and hence in the success of the exotics. However, we do not know at present whether the predators or the forest changes are the more important.

Others have reached the same conclusion by completely different means. After a lifetime's meticulous documentation of the changing fauna and flora of his Hawkes Bay sheepfarm, Guthrie Smith dismissed the competition theory, on the grounds that native birds were

> neither debarred from their fair share of food, nor intimidated by the presence of the newcomers . . . [on the contrary] I have seen the frail-looking fantail hawking nonchalantly for insects in a deluge that was killing the homestead sparrows, quail, pheasants and other aliens wholesale. . . . In these storms, species whose forebears have not been accustomed to face seven, 14, and 20 inches in three consecutive ceaseless days' rainfall, perish in great numbers. . . . The real cause of the diminution of native birds is easy to give . . . woodland species cannot live without woodland, jungle and swamp-haunting breeds cannot survive without jungle and swamp, they cannot feed on clover and breed on turf. At one time there were on Tutira many hundreds of acres alive with forest birds; not one single individual now exists on many of these localities, because not one single tree remains.[76]

More recently, C. L. McLay has shown that the number and diversity of native birds declines steadily with the proportion of introduced species, in a definite series from virgin native forest, through modified and exotic forest to suburban bush and gardens.[77] And out in the open pastures, the introduced skylark apparently does not compete with the native pipit for food, since skylarks eat mostly seeds and pipits eat mostly insects – at least, they did on the sheep station at Tokomaru Bay (north of Gisborne) where A. S. Garrick studied them.[78]

We ought not to overlook the fact that some native landbirds have actually *increased* as the landscape changed from forest to farmland. One of the most spectacular increases was in the number of harrier hawks. In 1916, one enthusiastic sportsman and naturalist liberated £50 worth of partridges on his land in the Waikato. The next morning he wrote, 'my son found 8 of them killed, with a hawk on each; he shot a rabbit, poisoned it with strychnine and secured 18 hawks with it. In driving from here to Tepapa [38 km] a few days ago I passed (within sight) 162 hawks, 17 of them in one bunch.'[79] These figures give some idea of the enormous abundance of hawks in the North Island at that time (they are less abundant now – like ferrets, they declined after the rabbits were finally brought under control); but alas, fewer native birds reacted to forest clearance by increasing than by decreasing.

There is also one whole group of native birds, the marine and shore-dwelling species, which have been relatively unaffected by the changes on land. Although they have been driven from their mainland nesting sites, most have been able to find at least some secure alternative places off-shore, where they can still raise their young, even if in smaller numbers. At any rate, far fewer endemic seabirds (four per cent) than landbirds (15 per cent) have suffered extinction in European times:[80] their contribution to the saga of the immigrant killers of New Zealand is tame by comparison with that of the landbirds.

Conclusion

It ought to be possible to disentangle the effects of the various environmental changes that have taken place in the last hundred years, or at least, to decide how important the predators were in comparison with the rest. After all, by 1884 there already were established museums and learned societies, such as the New Zealand Institute (forerunner of the modern Royal Society of New Zealand), many observant scientists and naturalists out in the field, and all the necessary machinery for recording their notes and publishing their papers and lectures. Unfortunately, the picture is, if anything, less clear than for any previous period, mainly because so many changes were taking place, and no one observer could take account of them all. Moreover, the two main islands had completely different histories, and by 1884 quite different faunas, each subject to the effects of different kinds of environmental upheavals.

In the North Island, almost all the most vulnerable resident species (the tuatara and kiore, and certain birds and insects) had already been exterminated, largely by predators, in Polynesian and early European times, and the massive deforestation of the last hundred years has not *so far* resulted in many new extinctions. However, the main blocks of North Island forest have been swept away over a very short period, and most of what remains is subdivided into small patches and invaded by introduced browsing mammals and exotic birds. Such conditions are less than ideal for many of the more sensitive native species that still survive in them, so perhaps the process of extinctions has not yet caught up with the process of habitat destruction: anyway the arrival of the mustelids seems to have had rather little effect by comparison.

The only part of the country where these processes were not very far advanced by the late 1800s were Fiordland and Westland, and there, large populations of vulnerable ground-dwelling birds were still present when the mustelids and ship rats arrived. The circumstantial evidence linking the disappearance of the South Island thrush, kokako, saddleback, bushwren, western weka, kakapo and kiore with that event is strong and very recent. So it is not surprising that the most pungent remarks on the effects of mustelids on bird life, and the most strident calls for their control, tend to have come from Fiordland and Westland.[81] To some extent they are justified: predation certainly could have been far more significant in the South than in the North Island during later European times, especially in the south and west – and it was witnessed by men like Douglas, Henry and Harper, and by oldtimers still living today.[h] However, events in the remote and south and west of the country loom large only because they are the most recent and the most vivid, and, just as we have seen before, the magnitude of the effect was not so much due to the undoubted prowess of the predators, but to the quirk of history that left the native wildlife of Fiordland and Westland – including many species unable to cope with anything more than occasional predation – more or less alone until the mustelids eventually arrived. Except in these rugged wildernesses, still relatively remote even to the present day, the effects of predators in the later European period have been trivial by comparison with man's sweeping transformation of the landscape.

Modern New Zealand is still justly famous as an unusually unspoilt, natural and beautiful country in an increasingly overcrowded, overdeveloped and polluted world. In our corner of the South Pacific, there are still lofty mountains, glittering snowfields, placid lakes, tumbling rivers, broad tussock grasslands and rich evergreen forest rolling down to the sea. There are still active volcanoes and glaciers, geysers and boiling mudpools, pounding surf and heart-stopping earthquakes. The air is still clear and the colours sharp, with none of the mildly romantic mistiness of rural England. No New Zealand rivers have suffered the stomach-turning fate of the River Mersey, which Liverpool undergraduates of my day used to call 'the rectum of England', although many New Zealand rivers have been tamed, harnessed or diverted for hydro schemes. But, however pure and magnificent the grand scenic wonders and the clean landscape of this lovely country, they do not altogether make up for the larger reality we have lost. There are still thousands and thousands of hectares of native forest that have never been logged, although many of their original inhabitants have gone. The neatly cropped, vivid green pastures dotted with sheep (more than 20 for every human New Zealander) are pleasing to the eye, except that their fences are stark wire strands offering no cover for wildlife, and there are hardly any patches of original forest left as oases in a green desert.[i] The military ranks of aromatic introduced pine trees have an appeal of their own, especially to the visitor familiar with the northern coniferous forests, but they are artificial monocultures of largely commercial interest, while millions of hectares of rich native forest, with all their inhabitants, have been swept off the map. These historical details do not spoil the beauty of contemporary New Zealand for the casual visitors, who see it only as it is now; but naturalists and historians know how different it was, only so short a time ago.

The modern rural landscape of New Zealand is clean, uncrowded and peaceful. But it is not natural. In this view over the Wairarapa towards the Tararua Mountains, all the dominant animals and plants to be seen are introduced: few of the original residents remain, even on the hills. Only 200 years ago this scene was probably more like the one at the foot of p.16. Author.

The first human colonists to reach New Zealand found a land filled with birds. This appealing painting, published in a 1948 children's book on Maori legends, illustrates the general appearance of the later 'classic' Maori warrior, and some of the birds that were common in his day. The profound changes that have taken place in New Zealand both before and since that period are emphasized by the absence from the painting of the many forest species extinguished in early Polynesian times, and the near or complete absence from contemporary forests of six of the 17 species in the painting – and also the original culture of the warrior himself. The birds shown are (clockwise from top) tui, fantail, kaka, N.Z. pigeon, (unidentified), pied tit, huia, saddleback, red-crowned parakeet, ? stitchbird, grey warbler, kakapo, N.I. kokako, weka, kiwi and morepork. A. H. & A. W. Reed.

5

PREDATORS AND THE CONSERVATION OF CONTEMPORARY WILDLIFE IN NEW ZEALAND

The difference between then and now

Opportunistic predators, like children, have no idea of self-restraint in the face of superabundance. On Christmas morning a six-year-old will tear the wrapping off every package in sight; working out the full potential of what is inside takes longer. Predators surrounded by abundant and vulnerable prey tend to become wasteful and fussy, killing more than they need and eating only the choicest bits, ignoring the rest. Catching the more cautious ones, the really important breeding stock, takes more time, effort and risk, so they put it off. Naturally, wherever such predators arrive, the worst that they can do is done very quickly, and only later do they settle down to the harder work of a more normal predatory life. In New Zealand, the native fauna that could not cope have now all gone – to extinction or, at the very least, to long-term exile on offshore islands. Their exodus means not only that the native fauna to be found on the mainland today is quite different from that described in Chapter 1, but also that the introduced predators[a] now find the hunting rather more difficult than did their forefathers.

Because there is an immense difference in conditions between the environments to which the alien predators were first introduced and the enrionments they live in now, we must make a very clear distinction between the role of predators in the historic extinctions and their role in determining the density and distribution of native birds now. Predators – human and animal – probably did play a very considerable role in the historic extinctions, but that does not mean that predator control is necessarily a good conservation policy today. The impact of predators upon their prey depends as much or even rather more on the characteristics of the prey as on those of the predator. The first predators to arrive found hunting absurdly easy, and their impact was great. Their modern descendants, deprived of the easiest game, find conditions tougher, the dice loaded against them, and their impact is accordingly less.

Unfortunately, few of even the most dedicated amateur naturalists in New Zealand appreciate the significance of the enormous difference between the historic and contemporary circumstances. They may know something of the tragic history of the native fauna and flora of these islands, and also that most of the historic losses are partly or largely the work of the introduced predators. However, few of them realize that, while it is one thing to feel deeply about the predators that exterminated whole species of birds in the past, it is quite another thing to translate those feelings into futile hostility directed against predators killing individual birds in the present. The depth of this animosity can be quite startling, as the following example shows. This incident arose as a result of a *bona fide* research project in Fiordland, officially supported and described in the Annual Report of the Fiordland National Park Board for 1979-80:

A live-trap for stoats is made from a wooden see-through tunnel, with a nest box attached at the side, which also acts as an anaesthetizing chamber. Stoats are so nervous, quick and well-armed (with teeth and stink-glands) that they can be handled only under anaesthetic. The bottle on the right contains anaesthetic ether, which is blown into the box through a small hole in its side. The ear-tag, and the pliers for inserting it, lie ready on the trap lid. Orongorongo Valley, December 1974. Author.

This young female stoat will stay under the anaesthetic for only a couple of minutes, so the ear-tag must be put in and all the necessary data recorded quickly. Ear-tags are the only way to give stoats an indisputable individual mark, which is very important when dispersing young stoats can travel across distances (e.g. 20 kilometres in five days[66]) that would not otherwise be believed. Author.

A . . . contract was granted in 1979 by the Board to cover [a research project which] involved live-trapping, ear-tagging and releasing large numbers of mostly juvenile stoats in the Eglinton and Hollyford Valleys. It is hoped that as many as possible of the 130 or so marked will be recovered over the next few years, in order to obtain a series of dead stoats of known age which have lived in the wild from the time they were marked as juveniles to the time they are trapped. A reward is being offered by the Board for the return of any marked stoat, providing one or both eartags are intact, and the date and place of capture are recorded. A reference collection of animals of known age is essential in working out means of determining the ages of stoats killed in standard traplines, which in turn is one of the first steps towards calculating the effects of control operations. Other useful information may turn up too, e.g. on the distance that young stoats may travel in search of a place to settle, and the proportion of marked young that are recovered in the first, second and subsequent years. The final results of this study will not therefore be available for some time.

The response was immediate. From the *Southland Times* readers' column:

20 June 1980

– I now have my thoughts confirmed when I read of the Fiordland National Park Board's involvement in the study of stoats in the Hollyford and Eglinton Valleys. Trapping, eartagging and releasing 130 stoats to me proves nothing. (You never hear of rabbit boards doing this to rabbits.) I presume these stoats have been doctored or muzzled, otherwise it is like trying to push the contents of a night cart up hill (you'll never win). Sir, are permits available? and would the reward from these marked stoats sustain a full yearly income as I am considering becoming a full-time

Professional Stoat Hunter, Te Anau

3 September 1980

– I have been told that recently 34 stoats were trapped, tagged and liberated in the Hollyford Valley. I have every reason to believe this is a fact. Only one voice was raised in protest and that was not the Royal Forest and Bird Protection Society which I think should have protested strongly.[b] Just what this act was supposed to achieve I cannot understand. If it was an attempt to find the movements of animals – well, I can tell them without moving. They head for the first bird they can get their teeth into. I have in a long lifetime prospected from Preservation to the Hackett River, just to the north of Awarua Bay, and there are stoats everywhere. Dick Thanery recorded stoats on Resolution Island before the turn of the century. I killed two stoats on St. Ann's Point and Anita Bay fifty years ago while prospecting in the Transit River country and in those days the Kiwi were very plentiful. I went back twelve years ago on a solo trip and found not a trace of the large kiwi population which I had seen in previous years. There is no doubt that the ground bird life will soon be a thing of the past and low-flying birds like fantails will pass away too. I once saw a stoat jump off the ground and catch a fantail on the wing and that bird was three feet above the ground. Perhaps, like the sole protester, I will be but a voice in the wilderness. It is possible to make enough noise to be taken notice of. I trust that this type of experiment will not be repeated.

Old Prospector

26 September 1980

– . . . I also see that the Park Board has liberated stoats in the Milford and Hollyford Valleys. This is the most silly thing anyone could have done. It won't be long before there is a campaign to eradicate our stoats and all our birdlife will be gone by then. What is wrong with our Royal Forest and Bird Society. I think they should be fined.

Joyce Hunter

The lower Hollyford Valley, scene of several field studies on stoats supported by the then Fiordland National Park Board. In Fiordland the final destruction of the last populations of native ground birds was recent enough to be witnessed by the parents and grandparents of people who still live there, and who vividly remember stories about the abundant bird life of 80-100 years ago. Hence, Fiordland people tend to be particularly concerned about control of stoats in National Parks. Author.

New Zealand Herald, 21 September 1980
DEAD STOATS WORTH $10

The Fiordland National Park Board is offering rewards for dead stoats with eartags. The Board is offering the $10 rewards as part of a major research programme into the ecology of the stoat. The senior ranger with the Board in Te Anau, Mr E. M. Atkinson, said yesterday that about 120 tagged stoats had been released to find out where they went, what they ate and how long they lived. He doubted, however, that anyone would get rich by catching stoats. 'I wouldn't try to make a quick buck out of it because a stoat moves very fast,' he said. He added that research so far had indicated that most stoats lived for only eighteen months at the most anyway. Once completed, the stoat research will provide some answers to how much of a menace the rodents[c] are in the forest and possible ways of controlling them.

Whangarei, 21 September 1980
Dear Mr Atkinson,

I was horrified to read the enclosed cutting in this morning's *Herald*. What fool dreamed up such a ridiculous venture? No wonder our forests and wildlife are in danger, when such asses are allowed free play with their hare-brained schemes. We all know cyanide, guns, traps, cats, stoats and weasels are responsible for the destruction of our wildlife. The $10 payout is quite safe. No one will catch a stoat. Undoubtedly the brainless ones[d] who released the stoats would not have sterilized them, and they will breed. For God's sake, wake up and behave responsibly. You don't deserve to remain employed.

Unpublished letter, correspondent's name withheld

Apart from the fact that many of these correspondents had got hold of the wrong end of the stick (stoats were not being *liberated* in Fiordland, we were just marking the ones that were already there) one can detect a note of righteous indignation in their letters which has at least something to do with universal and deeply buried ideas of moral behaviour, inappropriately generalized from human ethics to the animal kingdom. Even Charles Darwin himself, overcome with the delights of his first sight of the South American jungle in 1832, was not immune to them:

> One day he got down from his horse to watch a fight to the death between a *Pepsis* wasp and a large spider of the genus *Lycosa*. The wasp made a sudden dart from the air, thrust home its sting and then flew off. Though badly wounded, the spider was just able to crawl into a tuft of grass and hide, and for some time the wasp ranged back and forth unable to find it. When at last, by an involuntary movement, the spider gave itself away, the wasp came in for the kill with wonderful precision – two quick stings on the underside of the thorax. Then the victor alighted and began to drag the body away. Darwin did the irrational thing that most of us would have done; he drove the wasp away from its victim.[3]

Half the world away and nearly 150 years later, almost the same scene was enacted outside the Mount Cook Post Office. The tiny village of Mount Cook is little more than a huddle of houses overshadowed by lofty mountains, and surrounded by windswept tussock much favoured by rabbits. The Mount Cook National Park rangers go to some trouble to shoot or poison rabbits whenever they get the chance. Nevertheless, when the postmistress looked out and saw a rabbit being killed by a stoat, she did not applaud the predator's assistance to the ranger's control programme. She made no philosophical comment about the death of one animal in nature bringing life to another one. Just as Darwin had done, she did the impulsive, obvious thing – she rushed outside and saved the rabbit, sending the stoat away hungry.[4]

This compassion for the victim of a predator – regardless of whether or not the victim is a pest in its own right – is especially marked in New Zealand, where all predators are regarded as 'wrong' for another, far more important reason. Predators are, after all, designed to kill, and the picture of wild predators pursuing their natural lives in the place ordained for them can be regarded as right and proper, even romantic: 'Thou makest darkness, and it is night: wherein all the beasts of the forest do creep forth. The young lions roar after their prey, and seek their meat from God. The sun ariseth, they gather themselves together, and lay them down in their dens.'[5]

What enrages the average conservationist in New Zealand is that mammalian predators are *alien* intruders into our forests; that they were never 'meant' to be here: and that they roar after prey that are doubly defenceless, provided not by God but by man's own bungling and mismanagement. Exposing the native birds to cats and stoats can be seen as the animal equivalent of throwing Christians to the lions: a horrible, immoral, spectacle that should be stopped at once.

It is no coincidence that, as the letters quoted above amply illustrate, official and public interest in the control of predators, and in conservation generally, is especially strong in Fiordland and Westland. These are the two places in the whole of mainland New Zealand where the ancient fauna has survived longest and, with the arrival of ship rats and stoats in the 1890s, was destroyed most recently. The public reaction to the live-trapping study was guaranteed to be more indignant there than almost anywhere else in the country. It illustrates well that, wherever the memory of the destruction of the last remnants of the original native fauna is still so fresh, the common loathing of the introduced predators is likely to be particularly strong. But that feeling should not influence national policy.

The way we think about the problems of our world today, and the solutions we try, are determined more than we may realize by these deep-seated attitudes. They are dangerous precisely because they are so seldom hauled out into the daylight of reason and made to account for themselves. It is easy to blame the introduced predators for many of the past extinctions, but not very useful if it leads us to forget the other environmental changes that man has brought about, or to attempt to manage the remaining native species with inappropriate techniques. Rational management policies for the contemporary world must be based on more than an instinctive human condemnation of all animals that live by reddened teeth and claws. First we must understand how the contemporary predators work; what they can and cannot do, and what restraints limit their impact in nature. Then we must understand the ecology of the contemporary native species, especially what really controls their distribution and numbers. Regardless of what predators may have done in the past, predator control now can be a worthwhile policy *only* for those species, and at those times and places, where predation actually still does limit wild populations.

How predators work

Predators are living animals capable of responding to changes in their environments: so too are prey. In the serious game they play against each other, life is the prize for

This sign was put up by the Mount Aspiring National Park Board in the Haast Valley. The statement 'Today ground birds are rare' is unfortunately true, and the common assumption that stoats must be to blame is probably also at least partly true, at any rate in the south and west of the South Island. But it does not follow that stoats were so damaging throughout the country, nor does it mean that control of stoats in the Haast Valley is necessary now. Author.

both – for the predator, in the form of a meal, and for the prey, escape. As in, say, the Davis Cup, they play as individuals; but the results are significant for the populations they represent. In a long-established game the outcome is always difficult to predict, because it is influenced by the many different characteristics of the individual players, and by the particular circumstances of the encounter.

First, it depends on the number of predators in relation to the number of prey. Predators are always relatively scarce animals, less often seen than herbivores, simply because fresh meat is less abundant than grass and leaves. The density of stoats is normally low, so their impact would be limited nowadays, even if they lived entirely on birds. For example, in 40 hectares of forest in the Hollyford Valley, Fiordland, in summer there might be roughly 150 pairs of passerine birds (native and introduced species) plus their young,[6] 40 to 80 ship rats, and one stoat. Unselective predators such as rats take eggs or birds less often than stoats, but can be a much more serious hazard to birds, because there are so many more rats per hectare than stoats.

Second, it depends on the rate at which prey killed are replaced. Mice can breed rapidly, and in certain conditions are able to add recruits to their populations as fast or faster than stoats can remove them: mice can comprise the main food of stoats, and yet still increase in density. In such conditions, predation cannot 'control' the numbers of mice. Only when mice reach high density do they cease to add recruits to the resident population, and then stoats can cut down this non-renewable standing crop very quickly. In Fiordland in early 1980 (after the beech seedfall of 1979), stoats were apparently able to bring forward the autumn decline in mice by two or three months, but only because the mice removed were no longer being replaced.[7] Clearly, the rate of predation needed to make any impact on a rapidly reproducing, short-lived species is much greater than for a slowly reproducing, long-lived species.

The impact of a predator on its prey is usually variable and always extremely difficult to work out. This stoat's nest found at Birdling's Flat contained the remains of three skylarks and one pied stilt, but we cannot conclude from that that stoats control the populations of skylarks or stilts. The fates of populations cannot be gauged only from the fates of isolated individuals. B. M. Fitzgerald, DSIR Ecology Division.

Direct predation by stoats could have exterminated the native thrush and kokako of Westland in a few years, but seems unlikely ever to exterminate the rabbits in Otago. This is why generalizations concerning the 'impact' of predation are impossible: it may control one prey but not another in the same place, or one prey in one place but not in another.

Third, it depends on how active and efficient the searching predators are, and on the number of refuges available to the prey. Stoats are agile, active and fearless, climb and swim well, able to attack prey much larger than themselves, and because they can get into holes little larger than that of a mouse, they can enter almost all the likely refuges that their prey may try to hide in. On the other hand, searching costs energy, and stoats must eat a certain amount each day to live. As the prey decline and searching time increases, there will come a time when the stoats are not making a sufficient net energy gain each day. Then they must emigrate or die, although there may still be prey surviving, at very low density, which can breed up again as soon as the stoats have gone. This is the reason that natural predators can seldom exterminate their prey, except when they have the advantage of surprise attack upon unprepared and defenceless species unused to being hunted.

Fourth, it depends on the element of risk involved in making a kill. Some predators live on large prey that can be found easily but killed only with difficulty (e.g. lions on open savannah); others live on small prey that can be killed easily but take some effort to find (e.g. badgers picking up worms). Lions kill only when they have to, because killing large prey is risky and takes a lot of energy: stoats attacking adult rabbits have the same problems. Both lions and stoats may occasionally get the worst of it, if they misjudge the defences or the mood of their intended victim. A weka is fully capable of killing a stoat, and an enraged doe rabbit with young to protect is equally capable of chasing a stoat halfway across a paddock. So, naturally, predators tend to avoid high-risk prey, unless they are driven by hunger, or find one already at some sort of disadvantage. Conversely, when a stoat and a mouse meet face to face, the advantage is certainly with the stoat; but first it has to find the mouse, and in that the advantage is more often with the mouse.

So we can see why the problems of the scientist trying to work out the impact of predators on their prey are comparable to those of the manager of a chain of supermarkets trying to decide how to organize his stock. For both, there is no simple answer; it all depends on the circumstances.

The difference between the historical and contemporary situations for predators in New Zealand is easy to see. When they first arrived, the native flightless birds were like once-only 'loss leaders', such as bars of chocolate being offered at half price. The kill was easy, but the replacements limited, so the stocks vanished in no time. Since then, things have been more complicated. The shop has contained a lesser variety of goods, but those still there are nearly always available, for one of two reasons. There are the higher-priced luxuries that everyone would like but no one can afford except on lucky days; slow restocking is sufficient to keep them on the shelves. At the other end of the scale, there are the ordinary things, like bread and potatoes, which are the affordable but unexciting things that sell steadily and are therefore also found in most shops, and restocked frequently. If the local store temporarily runs out, shoppers can easily fill their needs from nearby.

In the forests and fields nowadays, large prey such as the hare and the native pigeon survive because, although the adults are highly desirable prey, they are not easy to catch, and their vulnerable young are hard to find and available over only part of the year. The easier prey, such as mice, rats, rabbits, wetas and small bush birds, are still around because their numbers turn over more rapidly. At times they may become scarce locally, but somewhere nearby there will be another population, a source of potential colonists and of food for predators in the meantime. But the details of when, where and how a particular local population will rise and fall are quite unpredictable.

Is predator control necessary in the mainland reserves today?

Control of abundant animal pests by man's own direct action is extremely expensive. For example, in the financial year 1979-80, the 60 regional Pest Destruction Boards in New Zealand spent $9.1 million, employing 679 people, mainly on the control of rabbits and possums on agricultural land. In addition, various local government officials undertook the same work in their areas, and the Forest Service spent an undisclosed amount on the control of browsing mammals in forests owned by the Crown (which include the National Parks).[8] The total spent on control of vertebrate pests that year probably passed $12 million. Control of predators on the same scale would greatly add to this burden; even if confined to important bird sanctuaries, the additional expenditure would still be massive. Clearly, no authority would recommend such action unless there was very clear evidence that it was necessary.

Effective control of browsing mammals in native forest has most gratifying effects. Plants are resilient and forgiving; in the right conditions the regeneration is profuse, and often contains species that were rare or absent before, provided there are sources of seed somewhere. An extreme example is in Hawaii Volcanoes National Park, where the commonest plant to spring up inside one small goat-exclosure was a previously unknown species.[9] In New Zealand since about 1965, when commercial venison recovery from helicopters became feasible, the numbers of deer in parts of the South Island have been much reduced,[10] resulting in convincing signs of improved regeneration of the vegetation. However, with the development of deer-farming[11] and the tightening of meat-inspection regulations, the recovery of thousands of dead deer per year from the hills has already given way to live-capture of hundreds, and presumably, if and when economics dictates, both will eventually stop. Conservationists may plead for subsidies to maintain the hunting pressure on the wild herds, but that would be justifiable only in certain highest-priority areas, such as the Murchison Mountains (p.144).

The relationship between deer and vegetation provides an instructive contrast to that between predators and birds. Browsing by deer can be thought of as a form of predation on the plants. If the soil is not badly damaged, the plants recover when the predation is relaxed, because their primary food resources – sunshine, air, water, minerals – are all still as abundant as ever. But birds are one step or more further up the food chain, and so are besieged from both sides; not only do they have to cope with predation, but also with disruption of their food supplies in highly modified or dissected forests. Hence there is no guarantee that they would recover if the predators could be removed. Whether or not they might, in any particular case, can

The advent of helicopter hunting in the late 1960s allowed hunters, for the first time, to remove deer from the hills faster than they could be replaced. When this picture was taken in 1973, dead deer were recovered and exported as venison; since then, the emphasis has been on capturing live deer to stock New Zealand deer farms. Author.

be determined only from studies of the food resources and population biology of the particular birds concerned.

The first approach to this question has been, naturally enough, to start with the predators. Unfortunately, even after some considerable research effort, there is still no firm information on the effect that any common predator, such as the stoat, has on bird populations in contemporary times. Stoats certainly do eat many birds – on average about 43 per cent of the 1250 stoats we examined, whose guts had any food in them at all, had recently eaten a bird. At Kaikoura, in an intensively-studied patch of bush in the South Island, stoats and weasels accounted for 77 per cent of known nest losses, affecting native and introduced birds equally, whereas rats and mice together were responsible for less than 20 per cent.[12] However, these facts alone are not enough to tell us whether or not predation by stoats has any effect on the present populations of the remaining mainland birds. To begin with, all feathers look much the same after they have been in a stoat's guts for a while, so we cannot tell what species of birds stoats eat – not even whether they are introduced or native. Not all species are equally vulnerable to the effects of population reduction, by predators or other means.[13]

A far more serious problem is that in the present day (as opposed to when the predators first arrived) the mere fact that an individual stoat A has eaten an individual bird B, or its young, does not tell us anything about whether total numbers of B will decline or not. If you are out walking in a National Park and you see a stoat galloping down a track with a fantail in its mouth, it is hard to avoid the impression

This is a sight that is guaranteed to enrage the average member of the Royal Forest and Bird Society. But in fact it tells us nothing about the impact of predation by stoats on fantails. The bird is dead, certainly, but that does not mean there will be fewer fantails in future if (a) the stoat found it dead, (b) the bird would not have survived to the next breeding season anyway, or (c) enough surplus fantails are produced each year to replace those killed by stoats that otherwise would have survived and bred. It takes a conscious effort to remember that this situation needs our concern much less than the wider issues of conservation (p.184). Cynthia Cass.

that predation by stoats is bad for fantails. The Boards of our National Parks sometimes get complaints from people who have had such an experience, followed up by demands for more stoat control. But visual impressions are not necessarily correct. Demands for control of stoats based on this sort of evidence promote only a 'bleeding hearts' attitude to conservation, i.e. one that focuses attention on the direct destruction of individuals.[14]

What we need is a conservation policy based on the ecological facts of how *populations* of both predators and their prey work in nature: we need a policy concerned with finding the factors determining the mean density of the population of the contemporary bird species, rather than with attempts to prevent mortality of individuals. If a long-term decline is caused by factors other than predation, the killing of some birds by predators need not alter the rate of this decline, and conversely, to prevent the decline it need not help to advocate predator control.[15] Regardless of what may have happened in the past, the size of the breeding population of that species of bird may now be decided by some aspect of the environment having nothing to do with predators, so that if stoats do not reduce the previous season's surplus to the maximum number that can breed, something else will. Because predation accounted for the death of a particular fantail, it does not follow that predation was the reason it died: some considerable proportion of all fantails die every year from causes other than predation, and this was so even in primeval New Zealand where carnivores were absent. The fact that a particular fantail died when a stoat caught it is of less interest than the reasons why the stoat was able to catch it. For example, if a stoat kills a fantail that is so debilitated by injury or hunger that it is unlikely to live much longer, the stoat is merely acting as a scavenger that does not have the patience to wait until the bird is dead.

The essential thing is to grasp that no conclusions about predation can be made from individual cases: the population is not necessarily reduced by the death of an individual, since the death of one may give another the space to live. Predation is a matter of relative numbers and rates of loss, which can be worked out only from studying whole populations of predators and their prey, simultaneously. Not every bird eaten by a stoat is a serious or preventable loss, since most of the contemporary common small birds produce more young each year than are needed to replace the adults that die. Nevertheless it is of course true that if stoats take too many, or if the number of surviving young produced each year is insufficient for other reasons, then predation can have a serious effect. How many is too many? Unfortunately, the answer to that question will be different for every species of bird, and for most of them we do not have enough information to say. The only way to find out whether the density or distribution of any bird species is affected by predation is to calculate, for each species in turn, the importance of mortality from predation compared with losses from other causes, comparing the normal variation in breeding success, distribution and abundance of the birds, in relation to their environmental requirements (allowing, for example, for forest damage) in similar places *with and without predators*. This is extremely time-consuming and laborious work, and has been done in adequate detail for only one of our native birds, the South Island robin.

Throughout most of the 1970s, J. D. Flack and his co-workers[16] observed the

The South Island robin (below) has been the subject of intensive study at Kaikoura by the Wildlife Service and others, largely because of its close relationship with the severely endangered Chatham Island black robin (above–a fledgling). Comparison of the life histories of robins on the mainland and on inshore islands has shown that robins react to predation on their nests by laying again. Wildlife Service scientists are now taking advantage of this trait to save the black robins by controlled cross-fostering.*/* Cynthia Cass.

effects of predation on breeding South Island robins at Kaikoura and on three offshore islands. The vegetation of all four areas is mainly *Leptospermum* (tea-tree) with patches of more mature and diverse forest. The mainland study area (240 hectrares) has ship rats, Norway rats, mice, weasels, stoats, ferrets, cats, hedgehogs and possums, plus three native avian predators, the morepork, harrier and New Zealand falcon. In the six years of the study (1971-6) analysed up to 1977, the robin population varied between 26 and 44 pairs, i.e. a density of between five and nine hectares/pair. On the most important of the three islands, Outer Chetwode (66 hectares) there are no mammalian predators at all, only moreporks, harriers, New Zealand falcons and wekas. The robin population is very dense, more than 60 breeding pairs (0.4 hectares/pair over the 21 hectares occupied by robins).

Studies so far have shown that the population biology of the robin on Outer Chetwode Island is conspicuously different from that at Kaikoura. The island robins occupied breeding territories of 0.2-0.6 hectares, much smaller than the 1-5 hectares at Kaikoura; they started breeding later and finished earlier; their productivity was always lower (modal clutch size two instead of three, one brood reared per pair per season instead of three, and 0.14-1.1 juveniles fledged per pair per season, compared with 3.0 at Kaikoura); nest failure due to predators averaged less than ten per cent, compared with 55 per cent; the annual adult mortality was about 17 per cent, compared with 23 to 37 per cent; and despite the much lower productivity of the island population, there were usually surplus non-breeding birds living in various marginal habitats awaiting the opportunity to find a breeding territory, whereas at Kaikoura, not all the available breeding habitat was occupied every season. The mainland population was also much less stable in numbers.

The climate and habitat at Kaikoura is as good or better than on windswept Outer Chetwode Island, so it seems likely that some or all of these differences in population biology and density are due to the absence of mammalian predators on the island. This suggests that effective predator control at Kaikoura *might* result in an increase in density of the robins there. On the other hand, if all other factors remain the same as when these data were recorded, it does not follow that predator control will be necessary to prevent the *extinction* of the Kaikoura population: the need for it, which might have been predicted when the predators first arrived at Kaikoura, and certainly still exists on Outer Chetwode (because it is a small island) has been overtaken at Kaikoura by the birds' ability to re-nest repeatedly, and the generally good survival of adults. That is not to say that they might not be made more vulnerable to other agents of extinction because their density is usually kept low by predators; but if they do become extinct at Kaikoura after some new environmental change, it will be difficult to know whether to blame the new change or the predators.

There are other islands which support higher densities of birds than can be found on the mainland, including several species which are now extinct there, but most of them also support various combinations of introduced mammals. For example, Great Barrier Island (28,500 hectares) has 21 species of native land and freshwater birds, including red-crowned parakeets, kokako, banded rail and kaka in especially good numbers, and the last substantial population of brown teal; it also has cats, rats, goats and wild pigs.[18] The steep slopes of Taranga (Hen Island, 476 hectares) support 12 common forest bird species, the only naturally surviving population of

North Island saddlebacks,[19] and kiore. Kapiti Island is relatively accessible and well-known for especially dense populations of breeding birds,[20] including several absent from adjacent mainland forests (e.g. kaka, little spotted kiwi, weka, parakeets, whiteheads and robins).[c] It also has possums, kiore and Norway rats. The large and diverse Little Barrier Island (3076 hectares) has all 19 native forest bird species that are widespread on the mainland, some in great abundance, plus the only natural population of stitchbirds;[19] it also has kiore and, until recently, cats. After intense effort (p.157), the cats were exterminated in 1980,[23] and since then, Wildlife Service officers report that predation on nesting seabirds has ceased; there is a noticeable increase in robins and parakeets; and the population of stitchbirds has multiplied sixfold. But it does not follow that such increased densities of any birds could be restored in mainland forests by predator control, unless all the most damaging introduced mammals could be removed simultaneously and permanently. This would perhaps be possible, at least in very small areas; but it would be outrageously expensive, and it is debatable whether or not the methods that would have to be used to achieve real reduction of all rodents, browsing mammals and carnivores over any substantial area might harm the birds as much as the alien invaders do.

Birds which are unable to cope with predation at all will be absent even from very large areas of their former homes that are still available to them. Some that have not been totally banished to offshore islands may still survive in less accessible mainland habitats. On predator-free islands in the Kermadec, Poor Knights and the Three Kings groups, spotless crakes live in all available habitats, including forests, but on the mainland they now live mainly in swamps. The standard Wildlife Service

The forest on Little Barrier Island has never been browsed by deer, goats or possums, and the only predators there now (since the cats were exterminated) are kiore. The numbers of some native birds there are astonishing. These tuis have gathered at a feeding trough near the resident ranger's house, though they are not held at artificially high density by the ranger's supplies of mainland honey. A. R. Thorpe, N.Z. Department of Lands and Survey.

technique of finding these secretive birds in thick vegetation is to play tape-recordings of their calls, to which the birds often respond in kind, or even in person. There are three recorded instances where the tapes attracted, not crakes, but mustelids, which undoubtedly expected to find spotless crakes after hearing their calls. Before the use of tapes to locate them, crakes were most often found as specimens brought in by cats. C. C. Ogle and J. Cheyne[24] have concluded that spotless crakes once did occupy forest on the mainland, as they still do on mammal-free islands; but now they tend to avoid predation by living only in vegetation standing in water.

Like most *post-hoc* explanations, this one remains to be proved, and it takes no account of other possible factors that might affect the distribution of crakes, such as what other habitats are available on the islands, and what other ground-feeding forest birds are present on the mainland. Moreover, swamps and wetlands are by no means out of bounds to predators, especially stoats and Norway rats, both of which are good swimmers. Guthrie-Smith, who recorded crakes in swamps and gorges at Tutira in the 1920s, found that by the 1940s they were gone. He wrote:

> where after floods on areas of drying mud there used to be revealed the imprints of the swamp and water crakes [sic], no little footmarks are now visible. . . . It is not man and settlement that, at any rate on Tutira, are responsible for these recent gaps and vacancies in its field life. Locally there still exists covert, locally there still remains food supply for the inconsiderable numbers of the missing species. The trouble is occasioned by the never-ending silent insidious furtive work of rats, cats, weasels and stoats.[25]

Conversely, several rare or declining species remain on Great Barrier Island in good numbers in highly modified habitat, but secure from mustelids and Norway rats. The picture is not at all clear: but even if it were like the proverbial crystal, our chances are practically nil of ever achieving predator control on the mainland on a sufficient scale to permit the most sensitive birds to re-occupy their former homes, even if the forests and swamps had remained unchanged in the meantime – which, of course, they have not.

On top of the 70 per cent reduction of total area of forest, and the effects of browsing by deer and goats, is the fact that the 6.2 million hectares of forest left (Table 1) is not all in one block, but in scattered patches of various sizes. This dissection of the remaining forests has serious consequences for birds. The remnant patches become equivalent to landbridge islands (i.e. islands once connected to a mainland) for forest birds, surrounded by hostile farmland, pine plantations or built-up areas where they cannot live. An analysis of the distribution of native forest birds in remnant forest patches by D. G. Dawson and K. R. Hackwell shows that (a) most species are found in large patches of forest, or in smaller patches close to a large patch; (b) most species prefer a low altitude (e.g., tui, pigeon) or migrate to shelter there in winter (e.g. grey warbler, parakeet); and (c) many are unable to live in modified (regenerating or partially logged) forests. They point out that forest reserves should therefore be large, unmodified, close to other reserves and include the full range of forest types and altitudes found in the region. Unfortunately, this requirement is not met in the present forest reserves.[26]

The effects of fragmentation of the once-continuous mainland forests into separate island reserves are drastic and, to some extent, predictable. Remnant

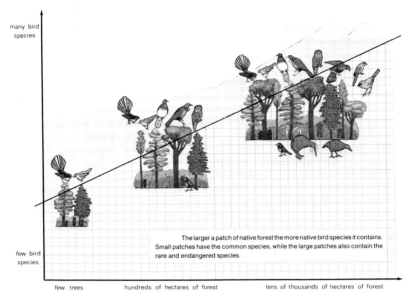

many bird
species

few bird
species

The larger a patch of native forest the more native bird species it contains.
Small patches have the common species, while the large patches also contain the
rare and endangered species.

few trees   hundreds of hectares of forest   tens of thousands of hectares of forest

Most of the native forest that remains is fragmented into scattered patches of various sizes. The lists of birds recorded in a large number of patches were analyzed, and the number of species plotted against the size of the patch. This diagram illustrates the result. The largest patches remaining include the cool-temperate beech forests on the mountain ranges, a fact which is emphasized by the mountain shown behind the large patch on the right. K. R. Hackwell, DSIR Ecology Division.

patches of forest generally contain fewer native species than a patch of equivalent size of the original forest, and more and more species are lost as the remnant islands become smaller. The bird lists for the remnant forest islands of a range of sizes form, like saucepans, a nested set; that is, virtually all the species that have survived in a small forest island are also present in the next largest, and so on through the series. The common small bush birds, the fantails, silvereyes, warblers, pigeons, tui and moreporks are found in relatively small forest patches of a few tens of hectares; larger patches (hundreds of hectares) have these birds plus additional species such as yellowheads, weka and robins; the largest (tens of thousands of hectares) have all these and also the rarities such as parakeets, kaka and kokako. The provisional Atlas of Bird Distribution in New Zealand[27] shows that forest species that were once widespread, such as kiwi, blue duck, weka, kaka, parakeet, robin, yellowhead, whitehead and kokako, are now absent from most small and isolated forest patches. This could happen only if the species were dropping out in a particular order as the forests diminish: as we might expect, extinction is not a random process. The first to go are the long-established endemic forest-dwellers (p.34) which are of greatest interest and in greatest need of protection, which cannot survive in small forest patches for long, and see the denuded landscape outside as an unsurmountable barrier prohibiting dispersal to the next patch. Most of those that survive are the widespread, recently colonized species tolerant of habitat changes within and between reserves.

Remnant populations of native birds protected in isolated forest reserves are still not safe from local (or even total) extinction, for two reasons: free-ranging species such as the kaka may require a greater area than the reserve provides, and sedentary species such as the robin are unable to disperse across the hostile habitats between reserves. I. G. Crook[28] has pointed out that the kaka may have the misfortune to provide the outstanding example of what can happen to the value of a small reserve as the forest around it is milled. The lower Inangahua Valley, part of the extensive area of Westland which the Forest Service is considering for felling, at present has large populations of kaka. One proposed Ecological Reserve in that valley has been extended to take in an area of apparently rich kaka habitat, in the hope that at least one locality with a high density of kaka may be preserved. However, kaka probably range over a much larger area than the proposed reserve, and so, as the forests around it are felled, the number of kaka using it will decline, and we may end up preserving only an area *that used to have* a large population of kaka. The robin's problem is the opposite: they range over very small areas in native forest only, and completely avoid open country.[16] Robins are not at present endangered, but the local populations in remnant patches are isolated from each other, and natural local extinctions cannot be corrected by natural dispersal. There are now large areas of habitat which, although suitable for robins, cannot be reached by them.

These two birds illustrate in different ways one of the first basic rules of conservation biology: conservation of species is conservation of habitats. As C. Imboden put it, 'The continued survival of the full genetic heritage of any species is not possible outside the habitat to which it has become adapted as the result of long evolutionary processes'.[29] To the extent that we allow the remaining forests to be dissected into small bits, unrepresentative of the original forests and unable to

The kaka, a large brown forest parrot, is an example of a native species which cannot survive in small patches of forest. Foresters may try to ensure that patches of good kaka habitat are left after milling, but if these 'reserves' are not big enough to serve the birds' needs all the year round, they will not prevent local extinction of kaka.
Cynthia Cass.

support many bird species each, we will have failed to conserve their habitat. Fortunately, we have at last realized the importance of applying the principles of proportional representation of habitats and biogeography to the design of forest reserves; but only just in time. The 1980 National Parks Act includes among its purposes the preservation, not only of areas that are scenically beautiful but also 'ecological systems . . . so *scientifically important* that their preservation is in the national interest' (my italics). Current discussions on planning reserves are certainly much better-informed than before, and the need to preserve large areas of lowland forest is well known (p.109). If that preservation is not achieved, it will be because of the way our society chooses to live now, regardless of the future. The problem is less to do with the wild life than with ourselves, and the solution must be not so much a biological, as a political one.[29] Logging the lowland forest for temporary financial gain must stop sooner or later: if later, when we know what the permanent consequences are likely to be, we need not expect our impoverished descendants to thank us.

The unwelcome effects of the dissection of forest into small patches are virtually inevitable: there is little chance that they could be prevented, or corrected, by predator control. There is ample evidence that predation is one of the reasons why many of New Zealand's most valuable and sensitive species are now confined to offshore islands. But the distribution of those that still survive on the mainland, and which have therefore shown that they can cope with predation to at least some degree, is determined largely by other factors such as their powers of dispersal, and especially the size of the areas of suitable habitat that they occupy. As these remaining patches get smaller, the resident bird species disappear one by one, and predator control cannot bring them back.

For example, soon after the First World War, the bush on a property in the East Cape region was cleared, leaving a small private reserve of seven hectares surrounded by farmland. At first, it supported at least four North Island kokako; then a single pair up to about 1934; and finally, one alone in 1935, which disappeared a few years later.[30] Although kokako are among the species which may be rather sensitive to predation, it is unlikely that predator control could have prevented for long their extinction in so small a refuge, or that they could have survived indefinitely on a predator-free island of the same size. The reason is that, although *individuals* might survive longer in the absence of predators, the long-term survival of a *population* depends not only on avoiding excessive mortality of adults, but also on producing viable new generations of young. The smaller the population isolated from contact with other populations, the greater the degree of inbreeding.

In theory, if the inbreeding rate is more than one or two per cent per generation, unfavourable genes become established in the line too rapidly, by a process known as genetic drift, and the stock begins to decline in vigour and fertility.[31] A fixed population of 50 animals will have lost about a quarter of its genetic variation after 20 or 30 generations, and along with it much of the vital capacity to adapt to future changes in its environment. The experience of animal breeders suggests that the expected number of generations to the extinction threshold is about 1.5 times the effective population size. Geneticists and conservationists have used these figures to calculate the minimum number of breeding pairs that must be retained if a species is to avoid that terrible doorway. There is no universal and generally valid number, of course, only a series of probability levels; and we usually lack enough data either to calculate these levels or to decide which to accept for a given species. But as a general rule of thumb, a commonly quoted estimate is 50 interbreeding adults for short-term survival, and 500 for longer-term security. We do not know if these figures are appropriate for New Zealand birds, but they provide a base to start from.

It is true that there are examples of species, such as the sea-otter, which have recovered from temporary periods of critically low numbers, as low or lower than these, with no apparent ill effects. But such cases do not hold out much hope for most of the seriously endangered New Zealand birds. First, there is no field evidence to show that, in a population so small that inbreeding is becoming a danger, the genetic hazard is any worse than the ecological disturbances that caused the decline in the first place. Once hunting had been stopped, sea-otters could return to home environments essentially unchanged in the meantime: New Zealand forest birds cannot. Secondly, inbreeding is less serious in species with higher reproductive rates. There are good reasons for cautious optimism for the black robin;[f] but most of New Zealand's other most seriously endangered birds are equilibrium species, which have the double disadvantage of inability to recover quickly from population reduction, and a greater likelihood of attendant long-term genetic problems.

In the present state of our knowledge, it is hard to escape the conclusion that control of predators in the mainland forests, though bound to be expensive, is not in itself guaranteed to have any beneficial effect in increasing the density or distribution of the non-endangered breeding birds. This may seem surprising, since we know that predators could have contributed to the decline of at least some of the species that have disappeared this century, and that now they are widespread and still eat many birds. But the reasons for this conclusion are simple. First, predation is only one of a whole suite of factors limiting bird populations, and for many species it may not even be very important, next to the effects of the dissection and destruction of habitat and food supplies, by clearing and by browsing and grazing mammals. Second, predator control work is effective only when intensive, and then not completely nor permanently; individuals of the non-endangered bush birds saved from predators will not necessarily live much longer, and other factors besides predation determine the breeding density of all birds, so the expense and time of ineffective or unnecessary control work is not worthwhile. Last, and most important, the effects of predators in the past cannot be reversed by predator control now. The historical and contemporary situations are totally different: the fauna of birds available to predators, and the structure of the forests, were entirely different a hundred years ago from what they are today. It is too late to slam the stable door: not only has the horse gone, but the stable itself has collapsed.

Predators and the threatened species

All this may be very well, but to the average naturalist there is at least one class of birds worth pulling out all the stops for – the endangered species. Surely there cannot be any argument about whether predator control is necessary to protect them – after all, they are the last remnants of the original native fauna that was decimated by predators in the past. Alas, even here the justification is a lot less obvious than it may appear at first glance. We must first establish whether effective predator control

The drastic impoverishment of the native fauna of New Zealand is best appreciated by archaeologists, who have first-hand acquaintance with the tremendous variety of extinct birds that no Europeans have ever seen alive. The difference between then and now is staggering. So is the difference between 100 years ago and now. The introduced predators are among the prime reasons for these changes, but predation is only one of the hazards facing the contemporary endangered species, and for most of them, not the most important. P. R. Millener, National Museum.

really could prevent any further extinctions. Is it really the best policy, and what effects might be expected, good or bad? Then we must ask; is control possible? What are the best methods to use? Does control really reduce the density of predators, or does it simply replace various forms of natural population control?

First, some definitions. The word 'control' means the artificial reduction in numbers of a species, temporary or permanent, for the benefit of some other species. In this case, we are discussing control of predators for the protection of birds. The idea of 'management' is closely related but different: managed species are usually regarded as desirable in some way, and management aims to increase their numbers. Red deer illustrate the contrast: in Scotland they are managed as valuable game, while in New Zealand they are controlled (at least in the Takahe Area) as pests.

The words 'rare' and 'endangered' refer, in this context, to birds which once were much less rare and in no danger of extinction, like the kokako, saddlebacks or kakapo that only two centuries ago were widely distributed and relatively common. There are two other kinds of rarity which are not at all in the same class. Some birds are rare, but not endangered. Species such as the falcon are widely distributed but uncommon on the mainland, and found only as a visitor on all but the largest offshore islands. These birds always live at low densities, because they are well up the food chain, where energy supplies are diminishing. Species such as the white heron are proverbially rare in New Zealand, but the same species is very common in Australia and elsewhere in the world. The total New Zealand population of between 50 and 200 white herons[32] is about the same as, or not much more than, that of the takahe. But the white heron is not a rare bird in the sense that the takahe is. If the

Okarito herons disappeared, they could be replaced (if necessary with human help); if the takahe disappear, they are gone forever.

In the New Zealand region – the mainland and the inshore and outlying islands – there are now some 27 separate populations of native birds that are considered vulnerable, rare or endangered. Of these 27, 22 are either already confined to offshore islands where most predators are absent, or their best chances of survival are there (Table 7). This leaves only two vulnerable (the blue duck and the North Island kokako), and three endangered (the black stilt, the takahe, and the kakapo on Stewart Island – which, in the context of predator control, is large enough to be classed as a 'mainland') endemic mainland birds which could possibly be saved from extinction by control of predators.

The black stilt is a long-legged wading bird which nests on the ground on the open riverbeds and swamps of the Mackenzie basin, in inland Canterbury. It has been declining steadily since the turn of the century, but much more rapidly since 1973. In 1980, R. Pierce[33] reckoned that it would be beyond recall by the year 2000. He argues that the main reason for the continued losses is not so much the destruction of nesting habitat by hydro-electricity works (there is still suitable habitat unoccupied, although flooding is reducing it); nor hybridization with its close relative, the pied stilt (black stilts prefer by far to pair with other blacks, and observed mixed pairs are due to a shortage of partners for unmated blacks). The main reason is predation, mainly by feral cats, ferrets and Norway rats. The nests and young of black stilts are much more vulnerable to predation than those of the pied stilt, because the black prefers to nest on dry banks of streams and ponds, where the cats and ferrets hunt;

The black stilt has lived in New Zealand since the Ice Ages, perhaps a million years or so; its close relative, the pied stilt, arrived from Australia less than 200 years ago. Black stilt nests are more susceptible to predation than pied stilt nests, which not only means that more black clutches fail, but also that black adults whose mates have been killed, or are pairing for the first time, often cannot find black partners. Cynthia Cass.

fewer predators venture out into the swamps where most of the pied stilts nest. Also, in its long isolation in New Zealand, the black stilt has almost lost several anti-predator adaptations, such as the 'broken wing' distraction display by adults; and it has extended the fledging period, so that the young are vulnerable for longer. In the 1977-78 and 1978-79 seasons, only one chick fledged from a sample of 20 black stilt nests.

One way to keep ground predators away from the nests of birds that can fly is to build a fence around the nesting area. The fence is no obstacle to the adult birds flying to and from their nests, and it protects the eggs and chicks from ground predators while they are vulnerable, without preventing the young birds from leaving of their own accord as soon as they, too, can fly.[g] In the Mackenzie Country, funds from the Royal Forest and Bird Society have enabled volunteer labourers to build two large exclosures in prime stilt breeding habitat. In the 1982-83 season, which was generally disappointing for breeding endangered birds elsewhere in New Zealand, the stilts reared a record crop of 18 chicks, of which 11 came from the two exclosures[35] – and this despite terrible weather, including a bad snowfall in late October (in the milder 1983-84 season, 25 chicks flew). But fixed-site fences cannot always protect nesting sites on the shifting riverbeds of Canterbury, so the black stilts are still in considerable danger; but with constant watching and careful use of management techniques (such as cross-fostering, in which black stilt eggs are reared by pied stilt parents) they still have a fighting chance.

The shrill whistle of the blue duck could once be heard in fast-flowing streams in the hills and right down to sea-level in the more rugged districts of both main islands. Nowadays, they are restricted to remote mountain areas, where their nests are usually built in inaccessible places under overhanging banks and in caves near the water's edge.[36] When their ducklings are on the water the parents are constantly alert, and react aggressively to predators. Murray Williams described to me one incident he witnessed when a stoat, boulder-hopping across the stream bed towards a family group, was spotted by the drake, and rapidly driven off. In the early days of colonization, blue duck were easy targets for explorers and bushmen (p.77); but now that they are strictly protected, the main danger to the scattered remnants of this once-common torrent duck is probably not predation, but the impoverishment of their habitat. They feed on the small aquatic plants and animals that live on and under the stones of the river beds. Scouring-out and erosion of river channels, which washes these morsels away, has always been a natural event after infrequent massive rain storms; but the degradation of the forests and soils of the catchment areas of the high country rivers inhabited by blue duck ensure that it now happens more often than it used to. Moreover, the introduced trout have similar tastes in food and living places to blue duck,[37] and can probably hunt underwater more effectively.

The takahe is a large, robust flightless rail, once found throughout New Zealand.[38] Its bones have been found in swamps, middens and caves in both main islands, but by 1840 it was already confined to Fiordland, and was rare even there. There were no confirmed records of living takahe for 50 years after 1898, and the species was considered to be officially extinct. In 1948, however, a population of around 500 takahe was found living in the alpine tussock high up in the Murchison Mountains, west of Lake Te Anau.[39] Anxious not to 'lose' it again, the Wildlife Service

The blue duck was one of the staple items in the diet of the early explorers on the West Coast. They are now strictly protected from predation by man, and other predators probably pose them little threat. Their main problems now arise from human mismanagement of the forests and rivers they live in. Cynthia Cass.

immediately declared the area to be strictly protected; but the population continued to decline anyway. The total occupied range contracted, and the known number of birds dropped 60 per cent in the 30 years between 1948 and 1968, mostly in one catastrophic slump in the late 1960s. By 1970, only about 200 birds were left: by 1982, only 120.

In 1972, a Wildlife Service team under J. A. Mills began an intensive research effort to save the takahe, and one of the first discoveries they made was that introduced mammals cause takahe serious problems. The mammals concerned are not, as one might expect, predators, but deer. During the summer both red deer and takahe feed on various kinds of snow tussocks. Careful feeding studies have shown that both prefer, not only the same species of snow tussocks (*Chionochloa flavescens* and *C. pallens*), but also that both are able to choose the individual tussock plants which contain the most phosphorus, an essential nutrient for both species.[40] High populations of deer in the past have severely damaged parts of the alpine tussocklands in the Murchison Mountains, since deer graze more thoroughly than takahe. The first plants to be weakened by persistent overgrazing by deer are the ones with the highest nutrient content, which are also the ones most needed by the takahe. Deer also badly damage the understorey of the adjacent beech forests, where some takahe feed in winter. In severe weather this allows the ground to freeze, hindering the birds from digging for their main winter food, the rhizomes of a fern, *Hypolepis*.[41] Reduction of the competiton from deer has become one of the first practicable management proposals to arise out of the Wildlife Service's intensified

Takahe have specialized to feed on snow tussock, in fact on particular parts of particular species of snow-tussock. It is a high-bulk, low-quality diet, one which leaves them little to share with competing species, such as deer. Inadequate nutrition is probably the main threat to takahe now. Stoats will certainly oblige by mopping up underfed birds whenever they get the chance, but control of stoats in the Takahe area is not the main management priority there at present. J. A. Mills, N.Z. Wildlife Service.

effort to save the takahe, although of course it is not a popular one with hunters. In the mid-1970s some 600-800 deer a year were removed from the 650 square kilometres of the Murchison Mountains, greatly reducing their density,[42] and the condition of the tussock grasslands is improving.

There are some people who cannot agree with the assumption that in Fiordland, deer are more dispensable than takahe, and that therefore the interests of the deer-hunters should be overruled in favour of those of the birds. Their opposition is even more intense when it comes to the wapiti, the North American equivalent of the red deer, a very fine sporting animal potentially able to develop a splendid trophy head, and which is found only in the range of mountains next door to the takahe. Their argument is that there is no need to blame the deer: the takahe are doomed anyway, because they are being killed off by stoats. Is this true?

The number of takahe known to be surviving in the Murchison Mountains between 1972 and 1975 appeared to be fairly stable. In the main study areas used for research on feeding habits and nutritional requirements, there were 22 marked adult takahe in October 1976. By May 1977 there were only 13 remaining, although one of the nine missing was later found, with a new mate, some distance away. Only one was found dead, so seven were unaccounted for. The cause of their unexpected disappearance is not known, but it coincided with a period of particularly high numbers of stoats in the Eglinton and Hollyford Valleys (about 40 kilometres from the takahe study areas) between December 1976 and May 1977, following the good beech seedfall of 1976 and the consequent increase in mice.[43] During the period when the takahe seemed to be doing well (1972-5), stoats were at their normally low numbers. This coincidence hardly qualifies as evidence for predation by stoats on takahe, and supporting evidence is scarce (takahe disappeared from only part of the area studied, whereas stoats are probably widespread throughout). Nevertheless, we all waited with some apprehension for the results of the field observations after the next breeding season coinciding with high numbers of stoats, which was from December 1979 to May 1980. There was no decline in the number of marked adults that year.[44] The reason for the difference is unknown, and so is the vital question of what part, if any, stoats play in the population processes of takahe. But all the while we cannot exclude stoats from consideration, we must continue to regard them as at least potential threats, although lower on the list, at least in the short term, than competition from deer.

There are two reasons for this view. The first is that stoats have been present in the takahe area since the 1890s, whereas deer have been there only since the 1940s. If takahe had been declining since 1900 at the rate we know they have been since 1950, surely there would be none left by now. The second is that the danger from stoats *may* be especially acute after a heavy fall of beech mast[h] but it is not proven, and anyway that happens only every three or four years, while the danger from malnutrition and nesting failure is certain and is acute every year. There are some areas where established pairs always attempt to breed, and yet have never produced a living chick; J. A. Mills suspects that this is because there just is not enough nutritive value in the tussock in their territory. The Wildlife Service at present regards the removal of deer and the enrichment of the tussock grasslands as the most important tasks.[45] If they are right, control of stoats in some years is secondary to fertilization of the

Takahe Valley in a snowstorm, May 1972. The survival of takahe in such a severe climate is remarkable – or it would be, if they were not already well-adapted to it. Subfossil remains show that takahe were once found all over the country. One interpretation of this is that they are forest birds which have retreated to Fiordland to escape from man and other predators. But J. A. Mills suggests that they are really 'Pleistocene relicts' which lived on the lowlands during the ice ages and have retreated to the alpine zone to escape the returning forests. If this theory is true, the takahe should be excluded from the list of species believed to have been exterminated by the Polynesians. Author.

tussock and control of deer in *every* year, at least until the takahe reach a new population level. Neither of the two alternative strategies – transferring takahe to an offshore island, or maintaining them in captivity[i] – seems viable at present.

The North Island kokako, or blue-wattled crow, is certainly much more at risk from various forms of forest destruction than from any other single danger. Kokako are most abundant in the last remaining stands of dense native podocarp-broadleaved forest in the Bay of Plenty, Northland, King Country and the central North Island.[47] The area of these forests has been drastically reduced since the last war. Not all the felled forests contained kokako at the time of logging, but some certainly did: Pureora was the most famous case. In 1978, after strenuous protests by the more active members of the conservation movement, logging was temporarily halted in parts of Pureora State Forest by a three-year logging moratorium, 1978-81.[48] Compensation paid by the government to timber firms for broken contracts was reported to total about $7 million, although independent studies claimed that the actual loss to the affected companies was less than $1 million. The government refused to provide a breakdown of how the figure of $7 million was reached, and only two per cent of it was paid to the 32 families who were affected by the closure of the sawmills at Pureora. Many people doubted whether any bird could be worth such an outlay: and unfortunately, even now the future of the kokako remains in doubt unless some very large areas of forest can be permanently reserved and rehabilitated for them, which could be done only at further high and perhaps even more socially unacceptable cost.

The blue-wattled crow, or kokako, has survived in the warm temperate forests of the North Island for 200 years since the arrival of the first predators, 100 of them in company with both ship rats and stoats. Its close relative, the orange-wattled crow of the South Island beech forests, disappeared a few years after ship rats and stoats got there. The reason for the difference is unknown, but it might be something to do with either the sudden immense irruptions of stoats that follow massive seedfalls in beech forests (p.99) or the relatively recent arrival of possums and deer in the last forest strongholds of the North Island kokako, or perhaps both. Cynthia Cass.

This part of Pureora State Forest was identified as an area of outstanding value for native fauna by the Wildlife Service in 1971. Nevertheless, it was clear-felled and burnt, and planted with pine trees (foreground). Although kokako are legally protected wildlife, it is not illegal to destroy their habitat. But the incident aroused considerable public protest, and it is unlikely that the N.Z. Forest Service will fell any more state-owned forest that could be inhabited by kokako.[51] N.Z. Wildlife Service.

In the last seven years, Pureora has been made a State Forest Park, and the burnt stumps of the old forest are over-shadowed by vigorous young pine trees (these were planted in 1977, photographed in early 1984). The thick grass around the young trees supports more mice and rabbits, and therefore more stoats, than undisturbed forest elsewhere in the park: the best-studied group of kokako, in the Pikiariki Road block, live less than a kilometre away. Author.

The causes of the decline of the kokako this century certainly include both logging and predators. Yet kokako can still be found in good numbers in forests which were invaded by predators in the last century, so there are at least some circumstances in which they can cope with predation. Current research shows that browsing by possums, goats and deer can badly deplete the kokako's food supply; and this effect is additive to that of human disturbance, since it is worst in forests that are recovering from recent logging.[49] If the adult kokako are not well enough fed, they may not be fit enough to attempt to breed: and those that do must often be discouraged, since predators destroy many nests every year. The known rate of production of young is frighteningly low: at Pureora, 15 pairs of adults produced only three fledglings in the seasons 1979-80 and 1980-81.[50]

Like the takahe, kokako may persist even after environmental changes have already exceeded the safety threshold, because both are long-lived species: we may see adults in the forests for years after their breeding rate has slipped below replacement level. If control of predators and browsing mammals could help them, it must be tried before the remaining birds get too old. Maybe it was only coincidence, but the first summer in which a really serious attempt was made to control predators throughout one 40-hectare kokako nesting area at Pureora (summer 1982-83) was also the first summer in which Forest Service scientists had the thrill of watching two kokako chicks grow from hatching, on 9 February, to partial independence by 15 May.[51] Only 27 kokako nests have been documented since 1885, but most of those protected against predators by poisoning and trapping have succeeded.[51] Unfortunately, that does not necessarily mean that the mainland kokako could be saved from extinction by predator control, because predators are not the only threats to their longterm security. But an effective management policy for kokako which has already reserved enough prime habitat, and expelled the deer, goats and possums from it, must then consider extensive seasonal predator control. Whether it is possible to achieve such a management programme over the large areas necessary, time alone will tell. The only alternative, which is equally risky, is to continue to transfer kokako to an off-shore island: if much more of their habitat is destroyed, that will be the only action possible.

The kakapo is a very large, mossy-green flightless parrot which, like the takahe, was once found in both main islands, but its range was already contracting well before the Europeans arrived.[52] At that time, kakapo were common in Westland and Fiordland (and, presumably, Stewart Island) but had almost gone from the North Island. Now there are none left on either of the two main islands except a few ageing males in Fiordland; on Stewart Island they are restricted to a relatively small area of 100 square kilometres in the southeast, where the last breeding population of kakapo in the world was discovered in 1977. By the 1981-82 season, more than 40 kakapo had been found by trained dogs, and 29 of them were banded. The total population then was reckoned at somewhere between 50 and 150 birds.[53]

But even as the birds were being found and marked, feral cats were removing them, almost as fast (p.74). In early 1981 the known mortality of kakapo was catastrophic. Of the 29 banded and released, eight were found killed (presumably by cats) and 13 disappeared. In the 1980-81 season, there were 49 occupied booming sites, on which the males, about every second year, go through their impressive

Modern technology has produced some wonderful aids for wildlife researchers in recent years. This kakapo is being fitted with a backpack containing a miniature transmitter. Tracking the kakapo moving about in thick bush at night would be impossible except by radio; so would finding out when and why they disappear. Radio work on Stewart Island has shown that the mortality of adult kakapo between 1980 and 1982 was very high, and due mostly to predation by cats. N.Z. Wildlife Service.

sexual display. By the 1982-83 season, only 11 of these sites were still occupied.[54] Excluding the birds transferred to other islands, the mortality of the wild males between the two years was estimated at a staggering 60 per cent, mostly during 1981. During the same period, nine of 15 kakapo fitted with radio transmitters were found dead, which also comes to a 60 per cent mortality in two years. These methods of calculation assume either that the same birds use the same booming sites each season, or that the kakapo carrying radio transmitters had the same chances of being killed as any other. Both assumptions could be wrong, but if they are not, the kakapo's situation is grim indeed.

Control of feral cats over the whole of Stewart Island (1746 square kilometres) would be an enormous task, and at the present rate of decline of the kakapo, could not be done in time. Control within the area occupied by kakapo is being done as far and as thoroughly as possible, but however great an effort is made, it is impossible to prevent new cats drifting in from the large uncontrolled area outside. Nevertheless, since the beginning of the intensive cat control work in 1981, the number of cats has been greatly reduced, and so has the number of kakapo lost; the known kill is down to one in the two years 1981-83. It is just as well; in August 1983 the Wildlife Service reckoned that the number of wild birds was down to 20 or less, of which only three were known to be females.[54]

If the kakapo is to survive, it is essential to protect these three birds from predators. But where? Richard Henry transported hundreds of kakapo to Resolution Island at the turn of the century, but none are known to survive there now. Seven kakapo transferred to Maud Island, in the Marlborough Sounds, had to be quickly rounded

Maud Island (foreground) was intended to be a predator-free refuge for endangered species. The shortest distance across the water to the mainland is on the other side of the island, and is about half the distance shown here. Stoats are perfectly capable of swimming that far, so island refuges need to be much further offshore than this to be safe from stoats. N.Z. Wildlife Service.

up and sent on to Little Barrier Island when a stoat appeared on the island in mid-1982.[j] They were joined by birds from Stewart Island, making a total of 22 kakapo on Little Barrier by August 1982. But one of them has since died, and several others have lost weight compared with Stewart Island birds at the same season, so it is not certain how well they are settling in. Such island transfers are not always successful; four kakapo and 19 little spotted kiwi taken to Little Barrier in the early years of this century failed to establish: and the number of other safe, predator-free and suitable islands available is strictly limited. We do not know enough about how to raise kakapo in captivity, and do not have enough birds to practise on. Even if more birds are found by new expeditions searching (again) other parts of Stewart Island, and new techniques to help them are worked out in time, the prospects are not good. So it seems distinctly possible that the kakapo may already be past the point of no return. The Wildlife Service can only go on doing what seems to be the right thing, and hope that some answer can be found within the lifetime of the last remaining birds.

Predator control in practice

We have seen that there is little evidence at present to justify any attempt to control predators, on a general scale, in average mainland bush country that does not contain any rare or endangered birds, and which has deer or possums as well as predators. On the other hand, we must be alert for evidence that might modify this conclusion; and in any case, there is no reason to go to the other extreme, and dismiss predator control as useless in all situations. Carefully planned and limited predator control could be worthwhile to help the black stilt, takahe, kokako and kakapo, and perhaps other species identified from future research. Unlike control of deer and possums, there is no user-group (e.g. stalkers or fur-trappers) who would oppose effective reduction of predators if it ever became possible. But the 'game-warden' approach has two great disadvantages. It would be terribly expensive, and it is probably not worth attempting unless it is done in association with rigorous protection or rehabilitation of the habitat. These problems are not confined to New Zealand: overseas experience is also that effective removal of predators is outrageously costly, and allows only short-term gains unless combined with habitat conservation.[58] But desperate situations call for desperate measures: it may be that we have no choice now but to face up to them.[59]

However, in New Zealand there is a further special condition: if rats are also present, control of carnivores (stoats and cats), even if possible, may not in fact be the best policy. This is not only because reduction of carnivores alone may not necessarily improve life for the birds very much: it is also because such a policy carries the risk of an undesirable side-effect. The idea that predators could control rats, or at least, abbreviate the large increases in rats that always follow good seedfalls, is hard to prove in mainland forests today. The statements of the early naturalists, such as G. M. Thomson, may well be true, but cannot now be confirmed: 'while rats are still very abundant, especially about the towns, there is no doubt that the spread of weasels throughout the country has vastly diminished their numbers, especially in the open'.[60] If the spread of mustelids had any effect on the rats, the weasel – the smallest of the three – would be less likely to have earned the credit than the stoat or ferret (p.100); and there is no way of telling whether it was predation, or the exhaustion of the most vulnerable food supplies, that diminished the number of rats in the late nineteenth century. The only comparable work on the mainland is merely suggestive. In the Orongorongo Valley, the density of ship rats reached as much as 37-49/hectare in 1950-51; but since then, many weekend huts have been built, and the number of feral cats in the valley has increased. Feral cats are now the most common predator there, and M. J. Daniel has suggested that predation by cats was sufficient to keep the rat population down to less than four per hectare during his study in 1966-69.[61] My study in Fiordland National Park in 1979-80 suggested that predation by stoats cut short the post-seedfall increase in mice and rats (p.127). But in neither case is there any way to work out what the density of rats would have been if predators had been absent.

Kiore on offshore islands without predators can reach very high densities,[62] as apparently they used to do in Polynesian times, before European predators arrived; but their disappearance from the mainland could be as much due to competition with European rodents as to predation by stoats or cats.[63] The decline in numbers of ship rats and Norway rats on the mainland since their first introduction is not doubted, and like kiore, they sometimes also reach higher densities on offshore islands than they do now on the mainland. Some authorities attribute these differences to the absence of predators on the islands, but again, other explanations are possible: for example, absence of competitors.[64]

Unfortunately, none of these studies provides conclusive proof of population control of rats by predators; but they are enough to suggest that there might be a substantial risk, especially in the podocarp-broadleaved forests where rats are most common, that effective control of stoats or cats alone could allow an increase in rats. Ship rats living in native forest now eat very few birds (p.72): but the historical evidence for the effects of ship rats on birds is so damning, that it would be unwise to ignore the possible consequences of even a temporary increase in their numbers, especially anywhere near a remnant population of endangered birds that might be especially vulnerable to rats (e.g., kokako in the North Island and kakapo on Stewart Island).

In places where these problems have been dealt with, and predator control is considered justified on ornithological and economic grounds, the next question is, what kind of control should be used?

There are two possible kinds: reduction of the overall population density of predators, and prevention of damage done by them.[2] Population control means the permanent reduction, by artificial means, of the average breeding stock of predators in a given area. Prevention of damage means the temporary reduction in numbers of predators only when and where they could be a particular threat, whether or not the breeding stock is affected. Population control is possibly only on relatively small islands: damage prevention is possible on the mainland, and justifiable in certain places. The benefits of both, if any, can be undone in a few weeks by immigration from nearby uncontrolled areas.

In order to achieve either kind of artificial control, and to avoid wasting money, it is clearly essential to have some idea of what controls the predator populations naturally. Recent research has given us some clues about the population dynamics of stoats, which are probably the commonest carnivore in the wild, and the subject of periodic protests from ornithologists and the public, especially after a good seed-year in the southern beech forests. Certainly, control programmes planned to protect the takahe and kokako must consider stoats, and therefore need this kind of information.

The normal mortality of stoats, especially of the very young, is extremely high. The potential productivity of females each year is also very high, but in most years it is counteracted by extensive mortality of the pre-independent young.[65] Whenever this juvenile mortality is reduced, a rapid population increase follows. In places where the food supply for breeding females in spring is fairly reliable, for example, where rabbits are always abundant, the female stoats produce some young every year, and the population density can be relatively stable. At the other extreme are the beech forests, where the food supply for the breeding female in spring is unreliable, allowing many young to survive one year and few the next, and so the population density of stoats in beech forests is especially variable. However, the mortality of the newly independent young is always very high, and the occasional population peaks decline swiftly, within a few months.[66]

Whichever kind of control is intended, it must start by reducing the local population of stoats. At present the only practicable means of doing this is by imposing additional mortality by trapping. Unfortunately, not only are stoats naturally short-lived, but their peculiar breeding cycle makes them very resistant to artificially augmented mortality. Both adult and young females mate in October or

The most efficient technique for killing stoats is the Fenn, a humane trap developed in England after the gin trap was banned there. The trap, opened out into a flat square, is set in a tunnel with the spring parallel to the sides. It should be sunk in the ground a little, and the large treadle covered with leaves. Author.

November, so 99 per cent of all females leaving the family group in January are already pregnant, each with eight to ten potential embryos in suspended animation. This means that, even if every male is killed during the summer and autumn, when trapping is most productive, the next generation is already assured. The early maturity, large potential litter size and prior mating of dispersing females are typical characteristics of 'opportunistic' species (p.33), and make them very difficult to control. For example, in the nineteenth century, English gamekeepers undertook the most intensive and sustained attempt at predation control ever made: 'equilibrium' species such as the pine marten and wildcat were exterminated in England, and clung on only in the Scottish highlands; but the 'opportunist' stoats and weasels were not permanently affected in numbers or distribution.[67]

It seems certain that effective population control of stoats, let alone total eradication, over any substantial mainland area by trapping alone, is impossible. The proportion of the population caught and killed has to be extremely high (probably over 80-90 per cent of the stoats present) in order to be greater than natural mortality.[2] The best we are likely to achieve in practice is about 50%.[66] Moreover, research on a variety of pest vertebrates, from pigeons to coyotes, has shown that mortality factors are usually compensatory, not additive.[68] In other words, if mortality due to trapping is less than natural mortality, it will probably only replace it: the trapper will be going to great lengths to kill stoats that are likely to die soon anyway. This simple principle has confounded very many well-intentioned attempts at predator or pest control. Worse still, female stoats must be caught in at least equal

A stoat caught in a correctly-set Fenn trap is killed almost immediately, usually by double fracture of the spine. This adult female had the distinction of being the first stoat to be caught by the first Fenn trapline in New Zealand, in Takahe Valley in May 1972. Pleased though we were with this success, later work has shown that trapping at any intensity ordinarily possible is unlikely to control the numbers of stoats in Fiordland.[66] J. A. Mills, N.Z. Wildlife Service.

numbers with males, although for at least half the year they are much less often caught than males.

A high capture rate can be achieved in a small area, with sufficient traps, but immigration constantly counteracts it unless the trapped-out area is completely isolated. But the areas where real population control of stoats might actually be worth the effort (for example, the Murchison Mountains and Pureora Forest) are large, and any real and permanent reduction in populations of stoats in large mainland areas seems unattainable with the knowledge we have at present; and any attempt to control stoats in smaller mainland areas, or those containing only the common birds, is probably not worthwhile. The best we can hope for is to try to reduce the amount of damage that stoats might do in the most sensitive areas during the nesting season, but even there control is not worth attempting during the rest of the year.

Some people will find it hard to accept that the possible benefits of predator control are really quite limited, even if it is effective – a large *if*. To them it seems beyond question that the only good stoat is a dead stoat, and that it must do some good to kill stoats whenever possible, even though many are left. Some people advocate the practice of taking a few traps along on every field trip and setting them around the camp,[69] or even laying poison in bush huts and deer carcasses. Knowing that stoats do eat very many birds, it seems obvious that, for every stoat killed, at least a few birds might avoid death for a little longer.

No one should be discouraged from helping with the urgent work of ensuring the wise conservation and repair of New Zealand's natural resources. But there is no point in putting valuable effort into the wrong policy. John Gould, a well-known naturalist, visited Australia in 1838-40, and later wrote ' . . . let me then urge them [the Australians] to bestir themselves, ere it be too late, to establish laws for the preservation of the large kangaroos, the emu, and the other conspicious indigenous animals: without some such protection, the remnant that is left will soon disappear, to be followed by unavailing regret for the apathy with which they had previously been regarded.'[70] Energy and legislation are certainly required, but predator control is not now a primary target. It is at least a hundred years too late to pass the one law that might have helped, the one prohibiting the importation of mustelids, which Buller hoped so desperately to see. Perhaps the embryonic conservation movement of Buller's day might have protected a decent population of kakapo long enough for it to have some chance now – perhaps Richard Henry's first attempts at translocation of kakapo to Resolution Island might have been hugely successful – if the odds against them had not included the stoat and the ferret. But we will never know, and it is pointless now to wonder. Instead, we should concentrate on the most important single issues at present, which are (a) the legal protection of as much as possible of the remaining lowland forests and wetlands, and (b) the continuing control of browsing mammals (especially red deer and possums) in existing protected areas supporting threatened species, especially the Murchison Mountains and Pureora Forest Park.

Next to these tasks, general population control of stoats in the average forest is relatively unimportant, at least until the other, more urgent management programmes have taken effect. Even though well-planned, limited control work in the takahe, kakapo and kokako areas is certainly justified, in the long term even this is less important than obtaining an indefinite guarantee of large, permanent and favourable breeding grounds. Predator control will not be necessary when the last aging takahe and kakapo have died without being able to rear chicks, and the last central North Island forests are reduced to blackened stumps or converted into pine plantations. Killing a few stoats whenever opportunity offers may give some personal satisfaction to the ardent stoat-hater, but will have no effect whatever on the general population density of stoats, nor on the level of predation by stoats on the common birds. The prospects of controlling the other predators that threaten to precipitate an extinction are not much better.

In the end, the fundamental problem of predator control is one of human perception. Take, as an analogy, a fly on a window-pane. If your eyes are focused on the fly, you cannot see the distant view beyond: to see that, you need to make a deliberate shift in focus, which can be felt as a physical sensation in your eyes. We need to make a comparable shift in intellectual focus when we are considering the problem of predator control. It is imperative to think of it as being an operation to protect the long-term prospects of a whole breeding population, a gene-pool. It may involve also the shorter-term aim of protecting the lives of individual birds; but it is possible to do that quite well and still fail to prevent the species from becoming extinct. For the rarest species most needing this and other forms of protection, the terrible dilemma is that only in the largest reserves is the gene-pool large enough to have any long-term chance of survival: and yet it is in just these large reserves that effective control of predators is most difficult.

The Snares are among our few entirely predator-free offshore islands. Until recently, fishermen have not been allowed to moor their boats to the island, because of the risk of rats getting ashore along the ropes. But it is hard to balance the case of the fugitive fauna and flora needing shelter from mainland predators, and the case of the fishermen needing shelter from the elements.[71] R.N.Z.A.F.

The most logical solution is, whenever possible, to avoid the problem altogether. The most vulnerable populations left are mostly on more or less predator-free offshore islands (Table 7), and it is worth almost any effort to make sure that their isolation is never broken. The trouble is, practically all the possible island refuges have problems of one kind or another. Little Barrier had wild cats, and a few years ago, many people were saying that it would be impossible to eradicate them. The area is relatively large, steep (to 772 metres), thickly forested and trackless. But the Wildlife Service proved all the pessimists wrong. The campaign to eradicate the cats started in 1978, and eventually involved 128 people, 27,000 poison baits, 75,000 trap nights, $11,178 worth of food supplies and the building of three huts and 67 kilometres of tracks. Altogether, 151 cats were killed, the last one on 23 June 1980.[23] The effort was magnificent, the story inspiring. Yet the island remains terribly vulnerable to fresh invasions, by accident or malice, just as years of work invested in preparing Maud Island as a predator-free haven for kakapo were undone overnight by the arrival of a single pregnant stoat. The Snares Islands provide not only a priceless rat-free refuge, but also one of the very few anchorages in a cold and stormy sea.[71] Here we must find some solution that is both humane to the fishermen, and also responsible towards the beleaguered Snares fauna.

The more our most critically endangered species become confined to the last few remnants of their former solitude that still exist, the greater the risk that, sooner or later, the outside world will break in and destroy them. The securest retreats are to some extent artificial, dependent on good management and on public goodwill; but with constant vigilance and reasonable luck, we should be able to keep them secure from predators for a little longer. It is not the ideal solution, but it is the best one we have left.

The evolution of the square-rigged sailing ship, and the expansion of trade around the world since about 1600 A.D., ended the secure isolation of many unique island birds. Wherever one of these rat-infested ships was careened, wrecked or even merely brought close inshore, the rats disembarked: other ships unloaded meat-hungry sailors or land-hungry settlers, whose effects were often just as disastrous. Alexander Turnbull Library.

6

INTRODUCED PREDATORS AND EXTINCTIONS
ON OTHER ISLANDS

The story of the decimation of New Zealand's native fauna is a sad one, but it is not by any means unique. Extinctions and invasions by animals and plants are perfectly natural, and there have been some spectacular examples of both during the long history of life on earth. There were mass extinctions at the end of the Permian period, about 225 million years ago, which wiped out fully half the existing major groups of marine organisms within a few ticks of the geological clock; and also at the end of the Cretaceous period, about 70 million years ago, which cleared the earth of about a quarter of the major groups living then, including the dominant and successful dinosaurs.[1] During the great 'faunal interchange' that followed the raising of the isthmus of Panama, a mere two or three million years ago, there were massive reciprocal invasions and mixing of the faunas of North and South America, and many families of endemic South America marsupials disappeared. S. J. Gould calls this 'the most devastating biological tragedy of recent times'.[2] Man had nothing to do with these events – they happened long before the human species had evolved.

So what is different nowadays? Simply that, natural invasions and mass extinctions were, in the pre-human world, rare. As man has extended his activities around the globe, so the numbers of new invasions and extinctions recorded each year have increased steadily. The correlation is not surprising, because man is by far the most widespread and mobile of all animals, and as he has explored the continents and oceans, he has helped other animals to spread too, usually to the disadvantage of the resident species visited. The host faunas most severely affected by these uninvited guests were nearly always those that had hitherto lived in secure isolation on islands, as the story of New Zealand illustrates very well. It has been estimated that, whereas less than 20 per cent of all birds are island forms,[3] some 93 per cent of the 175 species and subspecies of birds which have become extinct since 1600 AD lived on islands (Table 5). By contrast, only 11 continental species and two subspecies are known to have become extinct in the same period. Islands constitute less than seven per cent of the earth's surface, yet 53 per cent of all the bird species currently considered to be threatened are island endemics.[3] New Zealand obviously does not have the problem on its own, so before we reach any conclusions about our own little corner of the world, perhaps we should pause for a moment to consider what has happened in other places.

A small Pacific island: Lord Howe

Lord Howe Island and its surrounding islets are mere dots in the Tasman Sea, about 725 kilometres northeast of Sydney and rather less than twice that from the northern tip of New Zealand. Lord Howe itself has a land area of about 13 square kilometres, flat in the middle but with hills at one end and two massive, sheer-sided, flat-topped black basalt mountains reaching 865 and 762 metres high at the other end.[4] On a clear day these mountains can be seen from 100 kilometres away, but they are usually hidden by cloud. Nearby are several smaller islets and rocky pinnacles, and offshore is a shallow lagoon enclosed by a coral reef – at 31° 31'S, it is the most southerly coral in the world. The climate is mild except on the tops of the mountains, and most of the island was, and much still is, covered with forest.

The island was discovered in February 1788 by the British ship *Supply*, whose captain promptly claimed it for Britain; but for the next 50 years it apparently offered little use to its new owners, other than as an occasional stopping place for passing ships. The island showed no obvious signs of ever having been occupied by the ancient Polynesians, and no attempt was made to colonize under British rule until 1834. After a slow start, the human population increased to 40 in 1878, and is now about 150. At first the people made their living by providing supplies and hospitality for visiting whaling ships, and now they do the same for tourists.

The earliest known descriptions of the birds of Lord Howe Island come from the journals and logbooks of contemporary naval officers.[5] Arthur Bowes, ship's surgeon on the *Lady Penrhyn*, went ashore on 16 May, 1788, and afterwards described the great sport he and his companions had had in the woods among the birds. There were white gallinules and brown woodhens which were 'walking totally fearless & unconcern'd in all part round us, so that we had nothing more to do than to stand still a minute or two & knock down as many as we pleas'd wt. a short stick'. There were

Lord Howe Island from the air, looking south across Malabar Ridge towards the twin basalt peaks of Mt Lidgbird and Mt Gower. Ned's Beach, where the grounding of the Makambo *allowed ship rats to invade the island in 1918, lies behind the cliffs in the left foreground.* John Disney.

large fat pigeons which were so absurdly tame as to 'sit upon the branches of the trees till you might go & take them off with yr. hands or if the branch was so high on wh. they sat, they wd. at all times sit till you might knock them down'. The birds were in great numbers and so ignorant of their danger that 'if you throw'd at them & miss'd them, or even hit them without killing them, they never made the least attempt to fly away & indeed wd. only run a few yards from you & be as quiet & unconcern'd as if nothing had happen'd'.[6] Bowes' observations, and indeed even the words he uses to describe them, are practically the same as those of Cook at Dusky Sound in 1773 (p.76), of Darwin on the Galapagos in 1835 (p.35), and Wakefield in Queen Charlotte Sound in 1839 (p.77). In addition to these familiar kinds of land-birds there were seabirds 'in their thousands', and great numbers of fine turtles on the beaches; and in the woods, many smaller birds including parakeets, magpies and silvereyes. There were no ground predators, and only one species of native owl.

From descriptions such as these, the attraction of Lord Howe Island for passing ships is easily understood. In the days when techniques for preservation of food were primitive, revolting or non-existent, a mariner's interest in fresh meat could become understandably intense. When the meat was also free, unlicensed and easily collected, over-exploitation was inevitable. In May 1788, three months after the island was discovered, four ships called at the island together (including the *Lady Penrhyn*), and the captain of one recorded that enough birds were captured to serve his entire crew for three days. The slaughter in those early days must have been

The Lord Howe Island Gallinule was a flightless white rail rather similar in appearance to the pukeko. It was originally very abundant, until it was discovered by British sailors. Besides being large and meaty, it was also tame and fearless, and never discovered the sailor's intentions until too late. The crews of visiting ships (the island was uninhabited at the time) exterminated the gallinules in less than 50 years. Cynthia Cass.

terrible, and although the white gallinules, woodhens, pigeons and parakeets were originally very abundant, the island is small, and the total stocks limited. The white gallinule probably disappeared before the first settlers arrived; the pigeon was exterminated, largely by hunters, in about 1850; and the parakeet was finally shot out by the settlers about 1870, because it damaged their crops. The woodhen survives, in small numbers, to the present day, much as the New Zealand weka has outlived many of its former contemporaries that disappeared in the days of the moa-hunters.

Along with the human settlers came, naturally, other kinds of immigrants. About 1850 several cats were liberated from one of the whaling ships; house-mice were accidentally introduced about 1868; dogs, goats and pigs came to serve domestic purposes, but no doubt often escaped their master's control. But whatever additional disturbance the early animal settlers caused was not enough to add any more extinctions to the three already caused by man. This is largely because the first animal colonists did not include the arch-enemy of island birds, the rat, which had apparently never reached Lord Howe Island. It is not clear whether or not the Polynesian rat ever got there,[7] though it would be surprising if it had not, since it has reached every other corner of the Western Pacific. European rats must have been aboard the whaling ships which frequently visited the island; but as there was no harbour, all ships had to anchor some distance offshore, which must have reduced the rats' chances of reaching the land. Anyway, if any rats of either kind were present before 1918, they were few, excited no comment (no specimens were ever sent to any museum) and apparently had no effect on the birds.

In June 1918 this happy immunity was permanently destroyed when the steamship *Makambo* was grounded on Ned's Beach, on the northeast coast. The rats on the ship had plenty of time to scurry ashore, for nine days passed before the *Makambo* was refloated. In a few years rats (*R. rattus*) were everywhere, and 'the sequel to their coming makes doleful reading', as Hindwood put it.[7] He quotes a local naturalist saying that, before 1918, 'the forests of Lord Howe Island were joyous with the notes of myriads of birds, large and small and of many kinds. . . . Today, however, the ravages of rats . . . have made the note of a bird rare, and the sight of one . . . even rarer. Within two years this paradise of birds has become a wilderness, and the quietness of death reigns, where all was melody.' Hindwood added that there could scarcely be a greater contrast between this melancholy scene and Bowes' account of the island in its pristine days. Within a few years, five species of Lord Howe native birds disappeared; the flyeater, fantail, vinous-tinted blackbird, robust silvereye and the Lord Howe Island starling. Together with the three that had already gone, these five brought the list of extinctions to eight – half the 15 or 16 species of landbirds indigenous to the island. Since the human settlement was small and the natural vegetation largely undisturbed, these losses are pretty clearly the work of the alien predators – the visiting European sailors and the rats.

These were not, however, the only immigrant killers to reach the island. The settlers, horrified by the plagues of rats, attempted to control them by deliberately introducing predators. Almost a hundred owls of various kinds were sent to the island between 1922 and 1930. Naturally, the owls could not get rid of the rats, which are still there; but the combination of both owls and rats no doubt helped to make matters worse for the smaller birds. However, the surviving landbirds now seem to be holding their own, while the seabirds seem to be unaffected – all the original species still breed there in countless thousands.

A far-flung Pacific archipelago: the Hawaiian group

The Hawaiian chain of islands stretches for 2450 kilometres across the middle of the north Pacific Ocean. The oldest (over 27 million years[8]), lowest (to 12 metres) and most westerly island is Kure Atoll, which measures about one square kilometre and lies about 5000 kilometres from Tokyo. At the other end of the chain the youngest (less than a million years), tallest (to 4200 metres) and most easterly island is Hawaii itself, which measures over 10,000 square kilometres and lies about 3400 kilometres from San Francisco. The total area of land in the group is 16,638 square kilometres (that is, only about 15 per cent of the area of the North Island of New Zealand) and 64 per cent of that is the main island of Hawaii. Like Lord Howe, but unlike New Zealand, all the islands of the Hawaiian group are of entirely volcanic origin – on Hawaii the volcanoes are still very active – and none has ever been connected directly to a continent.

The Hawaiian islands have an amazing range of local climates and natural vegetation, from saturated rainforest on the windward sides exposed to the moist north-easterly tradewinds, to arid scrub on the leeward sides; from year-round subtropical warmth at sea level to winter snow on the tallest peaks of the main island.[9] Because of these contrasts, forest never did cover the whole land area; even in

On the main island of Hawaii very little native vegetation remains on the lower slopes, but at a thousand metres' altitude in the Hawaii Volcanoes National Park there is a form of warm-temperate forest, with abundant ferns and tree ferns, that reminds a New Zealander strongly of home. Many native birds are now confined to these higher-altitude forests and, as in New Zealand, those that could survive only in the drastically depleted lowland forests have almost or completely gone. Author.

primeval times there was no forest above about 2000 metres, and the driest areas supported only a low, open scrub. But everywhere else the forest was diverse and luxuriant, especially in the lowlands; and it included tall timber trees, shrubs laden with berries, and endless varieties of flowers, palms and ferns. It was filled with a great variety and abundance of birds – at least 81 endemic species spread in various combinations through the individual islands.[10] There were many kinds of small forest birds with brilliantly coloured plumage, and at least 13-15 species of flightless ground birds, including geese, rails and ibises. The avifauna of Hawaii, like that of New Zealand, evolved in a long-secure isolation and freedom from ground predators, although in pristine times there were three species of owls, a sea eagle and two hawks.

Polynesian voyagers reached the Hawaiian islands in about 400 AD. Like those that colonized New Zealand, they probably came from the Marquesas. They brought with them dogs, pigs, chickens, and various food plants including taro, yams, sweet

potatoes (kumara) and bananas.[11] The culture they established depended mainly on cultivation of these crops, and to a lesser extent also on meat from their domestic animals,[a] from the sea and, at first, from the forests. They also brought, probably as stowaways, Polynesian rats, geckos, skinks and various kinds of invertebrates and weeds.[12] Over the 1400 or so years that they had the islands to themselves, the Hawaiians developed a complex social and agricultural system, and, by about 1650, a population totalling between 200,000 and 250,000[11] – probably reaching or exceeding the capacity of the islands to support them.

To feed such numbers the Hawaiians had to clear the native vegetation, establishing permanent fields in the wetter areas and using shifting slash-and-burn methods in the drier forests. From the early European explorers' descriptions it is clear that, on the main islands at least, all the lower slopes of the mountains below about 440 metres (and in places up to 900 metres) supported only vegetation induced and controlled by man – fields and plantations where the rainfall permitted farming,

By the time Captain Cook arrived in Hawaii, land clearance by the indigenous Hawaiians was already very extensive. Forests that once would have reached the shoreline were driven back to the steeper mountains, so that Cook found he could nowhere get supplies of cut wood 'at any distance convenient to bring it from'.[13]
Alexander Turnbull Library

and extensive grasslands where it did not. The ingenuity of the Hawaiians in building irrigation works and fish ponds, and exploiting every available space for plantations, was often applauded by the early European explorers, even though modern evidence shows that the islands were already beginning to show signs of large-scale erosion.

The Hawaiian Islands were first made known to the western world by the indefatigable Cook in 1778. The European settlers that followed him came for a variety of reasons, but the consequences for the native inhabitants were nearly always bad. In the middle of a vast ocean an anchorage offering abundant supplies of food and water, and a congenial climate for rest and recreation, attracted European and American ships like flies to a honeypot. Naturally, all sorts of undesirable western influences (including disease, alcohol and imperialistic pretensions) came with them. From 1778 to 1803, missionaries and settlers introduced European farming methods and stock, including cattle, horses, donkeys, sheep, goats, rabbits and English pigs.[14] Only the pigs were not a new experience for the islands, but they produced better pork and so in time largely replaced the Asian pigs of the Hawaiians. Escapees of all these domestic species became wild, and feral herds of cattle and goats numbering in the thousands were causing extensive damage to the forests by the 1850s.[15]

Those universal commensals, the mice, cats and Norway rats which have accompanied man to all corners of the globe, no doubt arrived with the very first ships; but they could not get ashore immediately because, until a wharf was built at Honolulu in 1825, passengers and cargo could be landed only from small boats. Cats probably had little difficulty in securing an immediate passage to shore, but the rodents had to wait their chance. Cats were therefore found living wild in the forests long before mice (1816) or Norway rats (1835). Ship rats arrived later still, probably between 1870 and 1880, for the same reason that they were late in reaching New Zealand (p.71). Axis deer were introduced in 1868 and became a pest within 20 years, so fortunately no more game animals were brought for many years. The recent liberation of three new North American game mammals, the mouflon (1954), pronghorn (1959) and mule deer (1961), is explicable only when one recalls that the Hawaiian Islands have long been influenced by the United States, where sport hunting is a national religion.

The settlers set about establishing large-scale commercial plantations of sugar, coffee and pineapples on the rich volcanic soils of the older islands, displacing many Hawaiian crops, and yet more forest, in the process. Unfortunately, damage by Norway and Polynesian rats considerably reduced the harvest, especially of sugar cane. In Jamaica, where rats were causing similar problems well before 1870, the planters had already tried the age-old solution of introducing a predator. The small Indian mongoose was liberated in the Jamaican canefields in 1872, and first reports of its work were favourable. An article in the *Planters' Monthly* in 1883, discussing whether it would be wise to try the same in Hawaii, warned that:

> The introduction and complete naturalization of an animal possessing such strong predatory habits and remarkable powers of reproduction as the Mungoose must have an important influence on all indigenous and introduced animals capable of being affected by it . . . it would be important first to learn more of the nature of the creature, for they might prove an evil.[16]

But time, tide and commercial profits wait for no man, least of all for one who wants to hold them up while he learns about nature. In the very same year, September 1883, the first batch of 72 mongooses were brought to Hawaii.

At first the trick seemed to work; there was a period of about five years between 1883 and 1888 when rat damage to sugarcane was virtually nil. But the planters' relief was short-lived. By the early 1890s the recently-arrived ship rats were staging a full-scale irruption, as prolific species so often do when they arrive in a new habitat. They were seen climbing about in trees during the day, destroying or fouling fruit and berries.[17] These new rats – as would any rats in such numbers – proved more than a match for the mongooses, especially as mongooses are active by day and rats (except when at massive population density) by night. Needless to say, over the years the rats and the mongooses have come to terms with each other; all four are found together on the largest islands of the main chain, and, whilst canefield mongooses do live mainly on rats, damage to the sugarcane now has to be controlled by non-biological methods.

The impact of these events on the native animals and plants of the Hawaiian islands was, of course, catastrophic. Until very recently, people have always assumed that most of the damage was done by the colonizing Europeans – in contrast to the indigenous inhabitants who were visualized, in line with the popular romantic notion, as living in symbiotic harmony with nature. Recent archaeological evidence has shown that nothing could be further from the truth. The impact of the Polynesians on the islands was at least as severe as that of the Europeans, although for different reasons.

Like all human populations, the pre-historic Polynesian inhabitants of the

The small Indian mongoose was introduced to the Hawaiian Islands in 1883 in order to control the hordes of rats in the sugar-cane plantations. Despite their weaselly faces, mongooses do not belong to the weasel family, the Mustelidae, but to a related family, the Viverridae. Most viverrids are small carnivores living in tropical and subtropical Africa and Asia, where they occupy the same niches as the mustelids of cooler regions. Gary Fellers.

Hawaiian Islands actively manipulated and modified their habitat, and it did the same to them. Their destruction of the lowland forests exterminated most species of endemic birds, land-snails and insects confined to that habitat, and drove others to inaccessible mountain refuges. Among the species that disappeared were all but one of the flightless ground birds, which Polynesian hunters (aided by dogs, pigs and rats) certainly killed for food. As usual it is difficult to disentangle the effects of forest destruction and predation, especially when both were a long time ago. But the combined result was that of *at least* [b]137 island populations of endemic land birds of 81 species, only 78 populations of 41 species were left by 1778; that is, no less than half of the original number of endemic bird species (excluding subspecies) were lost in the 1400 or so years between the settlement of the islands by Polynesians and the arrival of the Europeans.[18]

The conclusion that the Polynesians were responsible for this slaughter depends on whether there is evidence that they and the extinct birds did, at least for a short period, inhabit the islands together. More than enough of such evidence is provided by carbon-14 dating of the subfossil bones, and also by the finding of 'marker' species (those we know to have been introduced by the Polynesians, such as the Polynesian rat, lizards and certain land snails) in the same archaeological samples as the extinct birds.

In the two hundred years since 1778, the number of island populations of endemic land birds has fallen from 78 to 64, and the number of species from 41 to 29.[c] This much-publicized decline is certainly bad enough, though it is proportionately much less than that of pre-historic times. It happened on all the major islands, though not at exactly the same time on each; most of these extinctions were recorded between 1892 and about 1910, but none since 1932.[19] The list of possible causes is long and depressingly familiar.

Logging for sandalwood (especially between 1810 and 1830), and browsing by cattle, goats and axis deer rank as secondary factors, at least in the sense that the disturbance they created was different only in degree, rather than in kind, from that which the birds had already endured at the hands of the Polynesians. Introduced birds brought two new factors, both of somewhat uncertain significance: competition for food, and unfamiliar avian diseases such as the birdpox viruses and avian malaria. Native birds susceptible to malaria are now more common above 600 metres altitude, but whether the restriction is due to the presence of mosquitoes or the virtually complete disappearance of all native vegetation from the lower altitudes is not at all clear. Direct hunting and collecting by Europeans was probably limited by the difficult access and precipitous terrain on many islands.

Cats, mongooses, and Norway rats certainly played their part, especially when they encountered any of the tame, ground-dwelling birds; but they were not agile enough to be of any real concern to the smaller forest birds. The sudden slump in numbers of forest bird species between 1892 and 1910 coincides most closely with the arrival of our old friend the ship rat. For the main islands, the evidence against ship rats is largely circumstantial but, as put together by I. A. E. Atkinson, it is damning. On Midway Island, one of the chain of small, eroded volcanic remnants strewn north-westwards from the main islands, it is a matter of simple historic fact. The endemic rail of nearby Laysan Island (related to the spotless crake) lost all its natural

The nene or native Hawaiian goose was rescued from extinction by captive breeding, and the US Fish and Wildlife Service is now attempting to re-establish a wild population. One of the release areas is this abandoned ranch on Mauna Loa. Besides having to learn how to feed themselves in this unnatural habitat, the released birds are also at risk from mongooses, feral pigs and dogs. Author.

habitat by the mid-1920s, after someone tried to start a commercial rabbit farm on Laysan in 1903. The rails were transferred, in 1891 and 1910, to two tiny islets on Midway Atoll, where they survived well and were treated as pets by the U.S. sailors stationed on Midway during the last war. Inevitably, ship rats got ashore from the naval ships in early 1943, and soon reached huge numbers; the rail was extinct by mid-1944.[20]

Contrasts: Britain and Australia

The effects of human colonization on remote Pacific islands are well illustrated by the three archipelagoes we have discussed in this book – the small Lord Howe group, the middle-sized Hawaiian group, and the large New Zealand group.[d] Half the native landbirds of Lord Howe have gone in a mere 200 years; half the native landbirds of Hawaii disappeared in prehistoric times, and another 15 per cent since; comparable figures apply to New Zealand (Table 3) and to other famous examples such as the Mascarene Islands in the Indian Ocean, former home of the archetypical extinct bird, the dodo.[21] The peaceful isolation these places enjoyed before man arrived certainly made them terribly vulnerable, regardless of their size; yet there seems to be more to it than simply the fact that they were islands. The British Isles have been colonized by man for centuries longer than any of these, and have suffered far worse deforestation, and yet they still retain the very great majority of their original bird faunas. Australia, the 'Island Continent', was colonized by Europeans only a few years before New Zealand, and has been transformed by them with the same ruthless speed; and yet the native animals that bore the brunt of that invasion were mostly not birds, but mammals. Why did history treat these places so differently?

The land area of Britain is about the same as that of New Zealand, but Britain is a land-bridge island – that is, it lies on the continental shelf of Europe, isolated by a seaway a mere 32 kilometres wide and less than 60 metres deep. Most of the land was covered with ice during each of the glacial stages of the Pleistocene, and the rest was connected to Europe across the bed of the North Sea. All the temperate vegetation and most of the resident animals – including man – were expelled from all but southern-most England during each glaciation. They have re-colonized the islands together since the end of the last one, between about 18,000 and 8000 years ago, from their refuges in southern Europe.[22] Few British species have had time to become even slightly distinct from their continental relatives: even the characteristically darker red grouse is only an endemic subspecies of the willow grouse of Scandinavia. Including migrants and introductions, Britain has about 180 species of breeding landbirds and 42 species of land mammals.[23]

Mature woodland covered almost all of Britain for about 4000 years, from the peak of its postglacial recovery to its final destruction over the last 300 years, which has been especially rapid since the beginning of the Industrial Revolution. In the early 1970s, woodland covered only eight per cent of Britain's land surface (only just over a third as much as at present survives in New Zealand), and the modern rural landscape of Britain is almost wholly man-made. But the transformation from entirely natural to almost entirely artificial has been mercifully slow. It has taken at least the 1500 years since the early Anglo-Saxon settlements of the fifth century AD;

Romantic impressions of the 'timeless British countryside' are appealing but inaccurate. The landscape of Britain has changed even more extensively than has that of New Zealand – certainly, much less of Britain's original native forest cover is left. These Welsh hills surrounding Llyn Gwynant were once thickly wooded and supported populations of beavers, bears and wolves. The contrast with New Zealand is not in the extent, but in the speed of the transformation. Author.

and (in the words of R. K. Murton) 'until the turn of the present century it had been achieved without serious detriment to a full and varied fauna and flora'.[24] The losses of resident (not endemic) British birds are relatively few. The pelican disappeared from Britain between the Iron Age and the Dark Ages; the crane in 1590, and the spoonbill in 1667; nineteenth-century losses include the bittern and osprey (since returned), the bustard, avocet, and ruff. More different species of birds now breed in Britain than at any period since modern records began (which almost certainly means more than at any time ever) and, perhaps because of slow, long-term changes in climate, they include unusual numbers of new species of northern or American origin.[25] Only six species of mammals have been lost, mostly large ones exterminated by hunters before 1800.[26] Several smaller ones, such as the wild cat, pine marten and polecat, clung on through the heyday of game preservation and now, in the age of conservation, are recovering well.

This enviable record is certainly not because Britain has never had the experience of being overrun by introduced predators. The ship rat arrived sometime during the eleventh or twelfth century AD, and the Norway rat in about 1728-29. Their impact is not known, but probably small. The American mink, a sort of high-class, glossy version of the ferret, was brought to Britain in 1929. Live mink are valuable, and the farm-bred animals were never intended to be let out; but of course they got out anyway, and escaped animals have been seen in the countryside occasionally since the 1930s. Only in 1956 were they confirmed to be breeding in the wild, and from about

Farmbred American mink have escaped and established in the wild in Britain, just as they have in many other European countries. (This artificial 'pearl'-coated mink could be mistaken for an albino ferret, but they are quite different animals.) But mink have had far less impact on the native British fauna than introduced predators have had in New Zealand, for reasons to do with the vulnerability of the prey rather than the prowess of the predator. Author.

Riverkeepers and Ministry of Agriculture officials tried every trick in the book to exterminate the newly established wild mink in England in the 1960s. But although thousands of mink were killed, the spread of these beautiful but unwelcome animals could not even be halted, let alone reversed. Trapping intensive enough to affect the populations of versatile, opportunistic predators such as mink (or stoats) cannot usually be sustained for long.
John Reader (Observer), Camera Press, London.

1965 the increase and spread of mink accelerated sharply. The British authorities reacted slowly, and the official trapping campaign got started far too late (1962) to exterminate the newcomers, or even to prevent their spreading. After killing 7000 feral mink in the decade up to 1970, all control efforts were abandoned as hopeless, and mink are now perforce accepted as new members of the British fauna. In lowland Britain, the mink partially replaces the native polecat, which has been locally extinct since last century.[27]

The extraordinary thing about this invasion by what is, after all, a versatile and accomplished predator, is that there is no certain evidence that any native mammal or bird has been affected – not even the native otters, which began to decline in about 1957, some years before feral mink became common in the mid-1960s.[28] Some of the native mammals and birds are rarer now than they used to be, but not because of introduced predators, and all are now well protected by legislation and by intense public interest.

Australia is huge (7.7 million square kilometres) and can be considered to be an island only in the sense that it is entirely surrounded by water. Its very dry climate[e] has made it an inhospitable place for human settlement, but during the 50 million years of its isolation, since it drifted away from Antarctica, it has developed a rich diversity of animals and plants capable of living in arid conditions.[29] Many of them were ancient residents of Gondwanaland, protected by Australia's isolation from the rest of the world just as those in New Zealand were, although the two countries and their inhabitants have evolved along totally different paths. One of the main differences between them was that the native fauna of Australia included two groups of primitive mammals, the marsupials and the monotremes, which were either replaced elsewhere in the world by the placental (modern) mammals that evolved later, or perhaps never lived there. In their secure isolation the Australian marsupials evolved a tremendous diversity of forms,[30] ranging in size from three to 300 centimetres tall, and they have invaded most habitats and exploited every kind of food, often in ways similar to those used by the independently-evolved placental animals such as dogs, cats and mice.

The earliest Aboriginals reached Australia at least 40,000 years ago, when cool-temperate rainforests were still widespread.[31] At first alone, and later with their only imported animal companion, the dingo, they adapted to the increasing heat and aridity of postglacial times, and developed a successful nomadic hunter-gatherer culture appropriate to their harsh surroundings. Their population was necessarily sparse, and the effects of their activities – hunting by men and dingoes, and uncontrolled bushfires – uncertain but probably drastic.[f]

Europeans visited Australia at infrequent intervals from 1606 onwards, and began to settle in numbers from 1787 (though not, this time, on Cook's recommendation[32]). The usual sequence of events followed: accelerated habitat destruction, introduction of domestic stock and many exotic birds and mammals (including of course, cats,

The impact of Aboriginal hunters on the original fauna of Australia is difficult to distinguish from that of climatic change. Nevertheless, the post-glacial droughts did not eliminate all forest refuges over the whole continent, and the extinct large marsupials had survived previous dry periods, before the hunters arrived.[31]
Australian Information Service.

Norway and ship rats from earliest days; foxes for hunting from about 1845; and rabbits for meat in 1859 – *after* the foxes!), and large-scale shooting of 'pest' native species such as wedge-tailed eagles, emu and kangaroo. The effects of the Europeans are described by H. J. Frith as 'profound and uniformly disastrous';[33] but they have not worked out quite the same as in New Zealand. Frith lists only two species of parrots, three island subspecies of emu and the noisy scrub bird among the species at present believed to be extinct or endangered, and four other birds as having suffered gross reduction in range and numbers, out of a total of about 600 species of landbirds. By contrast, he lists a total of 32 species or subspecies of native mammals in the same categories, out of a total of 236 indigenous species. Most of the Australian mammals and birds that have declined or disappeared are forms dependent on the natural vegetation of the arid inland. Europeans, with fire, rabbits and the systematic imposition of an alien pastoral industry have exterminated many inconspicuous native species almost casually, often without ever realizing that they were there. But in a land already well supplied with predatory reptiles, native rodents and native marsupial carnivores, the impact of the cat, European rats and fox has been less significant than that of habitat destruction.[34] All but a tiny minority of the native birds have survived the arrival of introduced predators without anything like the trauma that native birds have suffered in New Zealand.

The dingo was brought to Australia (except Tasmania) by the Aboriginals, probably about 13,000 years ago. The equivalent native predator, the marsupial wolf or thylacine, was restricted to Tasmania by the time the Europeans arrived. This and other evidence suggests that the dingo displaced the thylacine on the mainland, probably because it filled the niche of medium-sized dog-like predator more efficiently. Australian Information Service.

Patterns of extinction

There are a number of intriguing parallels and contrasts between these stories and that of New Zealand, which together might lead us to some interesting conclusions about the effects of introduced predators on the native faunas of these countries in the past. In the next and final chapter, we will consider the implications that these conclusions might have for the present and future conservation of the remnants of our own particular native fauna.

(1) A common feature of all the Pacific Island extinctions, regardless of cause, is that not all birds react to disturbance in the same way. People talk about 'the native birds' of a place as if they were like a herd of dairy cows, a group of individuals with roughly similar characteristics, ancestry, needs and responses to environmental changes. In fact the native species of long-isolated islands range from the long-resident endemics, unlike anything else in the world, to the most recent colonists indistinguishable from their overseas relatives. Extinction does not act randomly among them; the most vulnerable are also always the most unique, the very ones that are impossible to replace.[g] For example, in the places in New Zealand visited by Cook, some birds he saw, such as the kokako, saddleback, bush wren, New Zealand thrush and New Zealand quail are now extinct; others, such as the bellbird, tui, tomtit, fantail, rifleman, kingfisher and cuckoo are still common. Even one hundred years ago field observers such as Reischek, Henry, Douglas and Harper often recorded ground-feeding birds in places where they are now long gone, as well as the small bush birds which we can still see today. The selective disappearance of the older, endemic ground-feeding birds during and since their time was already obvious to these pioneers, and Henry put his finger on one of the reasons:

> And just now the 'lords of creation' have imported ferrets and weasels that prey on all such things that sleep on the ground, and as *kakapos cannot be expected to learn in a day what their race had forgotten for thousands of years*, the chapter of their history is in all likelihood coming to a close[37]

(my italics). Likewise, the absurdly tame, ground-dwelling gallinules and rails of Lord Howe and the Hawaiian Islands were the first to disappear, while many of their smaller, quicker contemporaries survive today.

(2) Only on untouched, undisturbed islands can such extremely vulnerable birds evolve and survive. The more remote the island, the less the chance of invasion; but equally, the worse the results when it happens. The first predator to reach such a place is guaranteed to have a devastating effect, merely because of the birds' own susceptibility. Unlike Britain, the Pacific islands have been isolated long enough to develop many endemic forms, and men, rats and cats were nearly always the first to burst in upon them. The reason that so many more extinctions are attributed to Norway and ship rats, which are less carnivorous than the other predators deliberately introduced later, is a matter of opportunity, not of lethal powers. The same two species also invaded Britain and Australia with far less effect, because the native species were far less vulnerable to them.

(3) Man has deliberately carried exotic predators to each one of the five groups of islands discussed in this book, although not always for the same reasons or with the

same results. The mongooses in Hawaii and the owls on Lord Howe Island failed to control the hordes of rats accidentally introduced by settlers, just as weasels, stoats and ferrets failed to control the rabbits intentionally introduced to New Zealand. In the human world, people tend to turn against experts and politicians whose policies do not work, and bird-lovers have turned out plenty of bad press for these imported predators that had no effect on the pests they were supposed to control, accusing them of devastating the native birds instead. In fact, the imports probably did less harm than is usually believed, because in all three cases the deliberately introduced predators were brought in very late in the story, long after the most sensitive birds had already been seriously affected by earlier disturbances caused directly or indirectly by man. By contrast the effects on the native fauna of the fox in Australia, and the mink in Britain, are debatable,[38] but not obviously disastrous. The reasons for this difference are not clear, but probably include the fact that both countries already had a variety of native predators, including extinct forms somewhat similar to the newcomers.

(4) The consequences of human colonization have spread out across all three Pacific island groups in two tremendous shockwaves, each followed by a period of readjustment (in New Zealand we have distinguished three). Each wave was spearheaded by one or more different kinds of predators, which encountered a different selection of native fauna totally unused to their particular brand of predation. Immediately, the species most vulnerable to them disappeared, leaving only those still able to co-exist in some way with the latest set of invaders. Then, and only then, other environmental changes overtook predation on the list of unfavourable influences on what native species were left. In the first phase, human hunters were certainly or very probably responsible for most of the extinctions – the European sailors on Lord Howe, the Polynesians in Hawaii and New Zealand. They hunted for subsistence, and concentrated on the larger, easily captured, meaty species. Smaller predators and habitat disturbances helped, but were less significant. The second phase began with an invasion of European colonists, human and animal, and has continued to the present. At first, the major cause of disturbance in this phase was predation too, but it was soon overtaken by habitat destruction, still the major problem today. The human motives involved were, and still are, less direct; in the early days, at least, birds and their habitats were damaged out of ignorance or necessity, and more lately, from carelessness or indifference. But habitat destruction is a far more lethal weapon than direct predation, and it affects more species, especially when it is done rapidly and spurred on by over-population or for commercial reasons. The reason that so many more native species became extinct during the first phase than (so far) since then is a matter of vulnerability, and does not reflect the degree of change imposed.

(5) Because island ecosystems are alive and dynamic, they do not merely suffer changes, they also react to them. In all three Pacific groups, the spaces left by extinct species have been filled by others. In New Zealand, at least 32 species of birds, and probably more, disappeared from the North and South Islands during the Polynesian era, while many of the 12 species now resident but indistinguishable from

their Australian relatives may have arrived then. In European times, at least ten new species have arrived unaided and established themselves, mostly waterbirds (Table 6), while eight full species, mostly forest birds, have become certainly or probably extinct during the same period. This is a high turnover for large islands such as the New Zealand mainlands, and reflects how profoundly the changes brought by European man have altered the relative proportions of habitat available for different kinds of birds, in addition to the changes already wrought by the Polynesians.

In the Hawaiian group, the clearing of forest and the building of irrigation channels and ponds by the Hawaiians considerably extended the area of habitat suitable for coots, ducks, stilts and herons, and all five native species of freshwater birds found there now are undifferentiated recent arrivals. The short-eared owl has apparently colonized only since the Hawaiians provided it with open space and Polynesian rats as prey.[39] On Lord Howe Island, the loss of the eight extinct species has been offset by the arrival of eight new species that have successfully colonized from New Zealand and Australia.[40]

(6) Although the environmental modification caused by man in Britain and Europe is far more extensive, intensive and prolonged than in New Zealand, Hawaii or Australia, it has not caused massive extinctions. The difference is that the British fauna has not been isolated for long or by far; its few endemic birds are distinct only at subspecies level, and many temporarily displaced resident species have, since birds began to be protected in Britain, recolonized from European populations; the environmental changes were slow and allowed time for birds to adapt; and the value of the remaining forest in New Zealand for birds is reduced for other reasons (pp.110, 139).

The old and the new. The laughing owl represents the eight or so totally extinct endemic species or subspecies of New Zealand which could not adapt to the transformation of their environment by the European settlers (the owl might actually have benefited from the Polynesians, who provided a new prey, the kiore). The spur-winged plover represents the 10 self-introduced species which have arrived and settled in during the same period. The processes of nature are dynamic, and man does not so much destroy them as alter their courses. Cynthia Cass.

7

CONCLUSION

The settlement of New Zealand by Polynesians, and later by Europeans, set off many drastic changes in the environment of the native fauna. In various ways these changes all made the safe, reliable habitats, to which the native animals had adapted over time, suddenly less safe and reliable – and without giving them a chance to adjust. It was as if the Noah's Ark they were on had suddenly shrunk to a quarter of its former size, and had been driven by storms into icy, turbulent seas littered with icebergs, against which it could be holed and sunk at any moment. If such a catastrophic loss of security had been all that the native fauna had had to contend with, things would have been bad enough: some of the occupants of the Ark would certainly have been lost overboard anyway. But what made matters much worse was that the Ark not only became a distinctly less comfortable place to live in; it was also boarded by pirates, armed (literally) to the teeth.

In Chapters 2, 3 and 4 we saw that the introduced predators came in waves – first the Polynesian hunters and the kiore, then the European rats and cats that arrived by accident, and finally the stoats, weasels and ferrets that were brought in on purpose. These predators were certainly very damaging, but it is important to remember that predation was only one of three major factors[a] responsible for most of the losses of native species since pristine times, and also that the relative importance of the three has not been the same in every place, or constant over the years. A recent survey of the forests and wildlife of South Westland, in relation to their past history and future use, is a good example of the changing interaction of the three.[1]

The forests of the West Coast of the South Island contain the last large and continuous areas of lowland indigenous forest left in New Zealand. Because lowland forest is the most valuable, both to native birds and to sawmillers, the last ten years have seen a protracted, vigorous and highly public wrangle between conservation and forestry interests for possession of the remaining stands.[2] The report of the Wildlife Service on the wildlife of South Westland concludes that habitat loss, predation, and the effects of browsing by introduced mammals are the three most important causes of the disappearance of many of the native birds which were common in Westland only a hundred years ago. But the historical sequence of the three causes of disturbance is clearer in Westland than elsewhere in New Zealand.

The cool, wet forests of Westland were not much affected by Maori occupation, so they remained relatively undisturbed until the middle of the last century. After the discovery of gold at Okarito in the 1860s, there was a sudden increase in human population, plus the associated demand for timber and farmland. However, many of the birds now missing seem to have begun declining before the logging of forests and draining of swamps had proceeded far, and in the southern districts of Westland, land clearing was generally not extensive enough to explain the decline of so many birds. The Wildlife Service report concluded that the extinction or severe decline of many South Westland birds may be blamed almost solely on introduced mammalian

In Saltwater Forest, Westland, a Forest Service crew load rimu saw logs. The Service's proposals to fell large areas of native forest stimulated much public discussion throughout the 1970s, and gave financial wings to field research, necessary in any case, which might not have been done so quickly otherwise. J. H. Johns, N.Z. Forest Service.

predators (including, of course, man). It is only now, after most of the damage within the capacity of predators has been done, that loss of habitat has become the most significant cause of *further* losses of native species.

Westland has repeated in miniature the history of New Zealand in general; the first wave of immigrant killers did their work with teeth, claws and stone spears, but they have long since given way to the most calculated and ruthless killer of all – civilized man. Ironically, our own species is the only one of all the introduced predators with the intelligence to realize that it is possible to slow the present massive acceleration in the rate of contemporary extinctions above the natural, evolutionarily inevitable rate; yet we do next to nothing about it. This is largely because the modern concept of conservation is recent and still not universally accepted; nothing like it was known in the past, and there are many places in which nothing like it is actually practised today.

Few of the early colonists, either Polynesian or European, had the concern for the welfare of wildlife that many modern westerners have suddenly developed. We now live in comfortable houses, with assured supplies of warmth, food and entertainment that they never dreamed of, and we have time and energy to spend on the protection of wildlife for its own sake. Conservation in this sense is a luxury that only the most secure of modern national budgets can afford; for most of the early colonists of both races, personal survival was the pressing need, and exploitation of nature was the only way to ensure it. There are only two, rather minor, differences between them and us. First, they had to exploit their surroundings with their own hands, whereas we exploit indirectly, by machinery and out of sight of our personal sensibilities, both our own resources and those of less-developed countries overseas.[3] Second, they knew less about ecological processes than we do, and were less able to predict the results of their actions than we are. They exploited the land and its resources out of honest ignorance and personal necessity; we continue to exploit, every bit as ruthlessly, with calculated and deliberate disregard for the consequences. So we are in no position to criticize the Polynesian and early European colonists who assaulted the natural landscape, each to the extent permitted by the technology they had. Moreover, there are parts of New Zealand where the exploitative mentality still thrives at least as vigorously in modern times as in the bad old days.

On the other hand, there seems rather little point in taking the more tolerant attitude criticized by C. A. Fleming: 'It seems we are reluctant to blame our fellow men for a pre-historic offence against modern conservation ideals, and would rather blame climate or the animals themselves.'[4] Neither criticism nor whitewashing are needed; just an acceptance of what has happened, and a determination to see that none of it happens again. The responsibility for changing our careless way of doing things is heavily on the present generation. By the time our children are old enough to take an interest in the arguments, it will be too late.

Conservation strategy for today and tomorrow

The story of New Zealand's land and wildlife strikes different people in different ways. The ardent conservationist tends to regard it as a shameful tale of exploitation and destruction by man behaving at his worst, and the extinction of many formerly common endemic species as an unrelieved tragedy. The detached zoologist might regard it more as a fascinating experiment in evolutionary biology, in the products of isolation and in the effects of invasion by alien species; he would see the extinctions as tragic only in that they reduce his opportunities to study evolutionary adaptation first-hand. The conservationist would like to restore the forests to their original state, if that were possible, or failing that, at least to the way they were before the Europeans arrived.[b] The zoologist would simply ask, 'Why?'

Somewhere between the extremes of the hotly emotional and the coldly rational responses, the various conservation authorities who have the legal responsibility to protect what is left must find a workable policy. From the information I have presented in this book, and in more detail elsewhere, I venture some suggestions concerning at least the part of the policy that deals with predators.

The first essential is to get to understand the targets of the policy, the predators themselves. Few animals are so universally misunderstood as they are. People tend to go by outside appearances, and to attribute human characteristics to animals. Predators that have appealing looks, such as bears or lions, have a good press image – to the extent that visitors driving through safari parks constantly ignore warnings to watch them only through closed windows. Less cuddly predators, such as stoats and rats, are regarded as personifications of evil, and their ability to control their normal prey is often grossly exaggerated. The fact that none of these misconceptions is true does not matter in most countries of the world, but in New Zealand it does matter.

Unless we can develop a more realistic and practical attitude towards the four-legged immigrant killers, and a more informed and critical attitude to the two-legged ones, which between them have so changed the natural life and landscape of New Zealand, we will continue to waste time with unnecessary arguments and inappropriate policies.

For example, it is true that some at least of the introduced predators have had, in the past, a significant impact on the original native fauna. Many people assume that, because the contemporary predators still eat birds, they still threaten the native avifauna and must therefore be controlled. This demand is based on two unstated assumptions; that the density and distribution of birds is now controlled only by predation, and that widespread effective control of predators is possible. Neither assumption is necessarily valid now (pp.129-57), so neither is much of a basis for a contemporary conservation strategy.

A rational policy for predator control must look beyond the individual case – the pathetic sight of the little feathered life snuffed out in the jaws of the vicious killer. The 'bleeding heart' attitude to conservation (p.131) does little good in the long term: much more helpful is the more demanding and rigorous discipline of modern conservation science. That tells us that the fate of a group cannot be judged from that of an individual without masses of additional information about the whole population of prey and the community it lives in. The most obvious data to start with, the analysis of samples of the foods eaten by individual predators, are actually the least informative.[5] Historical data, though full of pitfalls, are much more interesting.

The histories of New Zealand and the other islands discussed in Chapter 6 demonstrate pretty clearly that, when predators first reach an undisturbed island, their impact is immediate, selective, and damaging in direct proportion to the naivety of the native fauna, rather than to their own skills as predators. Within a few years the most vulnerable species will have disappeared, and control of the variety and abundance of the hardier species, that can cope with predation at least to some degree, will probably pass to other factors, particularly the distribution of their habitat and the abundance of their food supply. On the middle ground are the temporarily rare and endangered species, neither tender enough to disappear in the first onslaught nor hardy enough to survive indefinitely without help. These three classes of native birds – the extinct, the hardy and the endangered – need our concern in different ways.

The extinct species are, of course, long past rescue by any management policy, but they still have very great significance as object lessons for naturalists and historians. We can learn much from them about how *not* to manage natural habitats, and should make sure we do so, if only out of motives of enlightened self-interest. Man has precipitated recent environmental changes which have exposed many species to the winnowing effects of natural selection; but man, too, is an animal, and subject to the same biological rules as the dumb creation to which he thinks himself superior. Even in the twentieth century, man is not independent of his environment, nor exempt from the process that leads to extinction, which ultimately is the same for all animals: the irreversible increase of deaths over births. The same kinds of environmental changes that affect birds will in the long run affect us too except that, unlike birds, we cause them as well as suffer from them. It is worth studying the history of the extinct

Conservation involves giving natural processes a chance to carry on unmolested. It is concerned more with the long-term protection of whole evolving communities than with the fates of individuals or species within them. The needs of conservation are best served by a carefully planned system of reserves in which every natural habitat is represented. This stretch of land from the coast to the Paparoa Range, crossed by the Fox River, contains some outstanding lowland forest and karst landforms (both underrepresented in existing reserves): it ought permanently to be protected. J. H. Johns, N.Z. Forest Service.

birds of New Zealand not only for their own sakes, but for ours as well: the message is the same for both – adapt or perish. Man is 'intervening in the evolutionary process with all the impact of a major glaciation', as Myers put it so graphically[6]: but man has a good deal more personal interest in the outcome than the glaciers had.

The hardy species are the ones that have survived on the main islands for at least a hundred years in company with the whole range of predators and other habitat changes, and therefore are able to come to some sort of terms with them. The management policy appropriate for them is that of *conservation*, defined in a recent, thought-provoking textbook as denoting 'programmes for the long-term retention of natural communities under conditions which provide the potential for continuing evolution'.[7] What conditions do they require? Simply to be left alone in their natural habitat, and enough of it. In the long run, the continued survival of any species genotype is impossible outside the habitat to which it is adapted: conservation of species and of habitat are the same thing.

Species are not fixed and immutable: they live in changeable surroundings, and given time and space enough, those that are not already too specialized will change in response.[c] This capacity to change and adapt is one of the characteristics of living things that we should be most eager to protect. We can see examples of it happening before our very eyes, usually in subtle ways, and more often in small, opportunistic species such as stoats and sparrows, rather than in the large and famous endemic ones.[9] We still do not fully understand how this adaptation works, but there is no doubt that the best chance of protecting the full diversity of wild life is to maintain natural habitats, and this can, *must*, be done even before we fully comprehend the workings of the irreplaceable relationships between species and their habitats. If we can ensure the protection of adequate samples of the full range of natural ecosystems while they still contain viable populations of species which are not at present in difficulties, we can perhaps prevent these animals and plants from becoming the endangered species of the future. Predator control can make little contribution to this long-term work (p.129ff).

The endangered species are those that are unable to survive even in their own habitat without man's direct assistance – perhaps because their populations have already become too small or scattered to avoid inbreeding, or their habitat has changed in some detrimental way. The management policy appropriate for them is *preservation*, a programme 'providing for the maintenance of individuals or groups, but not for their evolutionary change'.[7] Most of New Zealand's famous endemic rarities belong in this group. In their day they were widespread and successful, but their slow way of doing things handicapped them when man and his imported animals turned their entire world upside down. If they are to survive, conservation of what is left of their habitat is not enough – preservation by active management (for example, fertilizing the tussock that the takahe feed on) is needed too. Predator control is not high on the list of management policies required for most of them, although some form of it is justifiable as part of the effort to prevent extinction of the takahe, the black stilt, the North Island kokako and the kakapo on Stewart Island – but only after all possible steps have already been taken to improve and safeguard the birds' habitat. Note that this suggestion is not based on any evidence that predators actually control the populations of these birds; it is based merely on the deduction

Preservation involves active interference in the normal course of nature. For example, stitchbirds were once widespread on the North Island, but by the 1880s the only population left was on Little Barrier Island. There they managed to survive almost a century of predation by feral cats, although saddlebacks did not. However, by the 1970s there were fewer than 500 stitchbirds left, and they would probably soon have disappeared, had not the Wildlife Service decided to exterminate the cats. Within a few years the population of stitchbirds increased to 3000, enough to supply colonists to other islands. Here, stitchbirds are being released on Hen Island. C. R. Veitch, N.Z. Wildlife Service.

that these are the only endangered species on the list that possibly could benefit from predator control (Table 7 and page 139ff). We may still be able to prevent extinction of these birds in the wild, provided we can secure their habitats first. Failing that, in the last resort they can be preserved on predator-free offshore islands, or in captivity, but such treatment is expensive, and does not always work (p.150).

Because the species most needing preservation are the large, well-known and appealing ones, the work of preservation gets more publicity than the equally valid work of conservation. But in scientific terms, preservation is only a short-term job, and in many ways is the less important of the two. Preservation is necessary only when conservation has failed. If the slowly dawning interest in conservation of late last century had advanced more quickly, or if the Maoris had retained more of their tribal lands intact, the list of species requiring preservation now might perhaps be shorter; likewise, the species we fail to conserve today will be the endangered species of tomorrow. By exceptional efforts and inspired management policies on offshore islands, some few species, such as the South Island saddleback, the black robin and the stitchbird, have been brought back a step or two from the brink of extinction over the last few years. But while attention has been focused on them and not on the greater importance of habitat conservation, the destruction of mainland forests and swamps has continued unchecked; so it is probable that in the course of the same few years many smaller, less appealing species have been pushed over the brink, some even before they were known to science.[10] Conservation is the prime task, and arguments for establishing reserves should be based primarily on the need to conserve whole ecosystems.

However, in our small and increasingly crowded world, proposals to establish new reserves are often hotly resisted, especially when the disputed land contains resources of commercial value, such as minerals or timber. The conservationist argument for the protection of whole and representative ecosystems for their own sakes is generally less compelling to the public mind that the preservationist argument for the protection of some particular rare species. Since the outcome of the wrangle is often powerfully influenced by the public's perception of the issues at stake, the greater but vaguer goal of conservation of natural habitat generally gets briefer media coverage than the lesser but more immediate goal of saving some particularly appealing native bird from irrevocable extinction.

Unfortunately, this can lead to the spurious argument that endangered birds are more important than the people who might also have good claim to the habitat in question; as P. F. Jenkins points out, to be forced into such a position is a strategic blunder.[11] It happened during the fight to save the West Taupo podocarp forests from milling, because the case for protecting the forest was based largely on the presence of a single species, the kokako. The result was that both sides could use emotive ammunition ('Save the kokako' or 'Sawmillers forced to leave jobs and homes') against the other side. Not only did this effectively hamper rational discussion, and polarized the debate, but also, if in the future the kokako do disappear from the disputed forests, the case for protecting that environment, with all its rich abundance of other life, may be diminished. Besides, such an approach also confuses the different aims of preservation and conservation.

There is no question of holding kokako in higher esteem than people, only that their needs are incompatible and cannot be served in the same place. Surely there is room for both wild species and for people, so that both can enjoy life in their separate ways, to the considerable enrichment of at least the human side of the bargain. On the other hand, although this concept may be perfectly true in a broad sense, the relevance of it is likely to escape the people with most to lose in a particular local dispute. The 'big brown eyes' approach has the powerful effect of concentrating public attention on the local issue, and, as it is very often successful, it will probably be used for as long as public opinion continues to be able to sway the outcome.[12]

The same prominence tends to be given to preservation when it comes to the allocation of funds for wildlife management and research. Since these are always inadequate, it would seem reasonable to channel them into projects with good long-term prospects, and avoid spending them on other projects with very little chance of success. The problem is that the most severely endangered species often develop a political or emotional appeal of their own, which is very difficult to resist; hence it is still considered justifiable to send out field expeditions to check up on reported sightings of species that are probably already extinct, such as the South Island kokako.[d] There are good practical reasons why endangered species management in New Zealand, as elsewhere in the world, has had to concentrate on the immediate cause of a species decline (e.g. predation) rather than the ultimate cause (e.g. loss of habitat) – in other words, on short-term preservation rather than on long-term conservation.[14] However, financial support of conservation, for the

*In the mainland National Parks, any increased financial support for conservation
would be better put into environmental education than into general predator control.
The weka still survives in Abel Tasman National Park, even despite the usual range
of mainland predators, and some individual birds at least are still as tame and
friendly as ever. Young people probably benefit more from an encounter such as this
than the weka would from any sustainable level of predator trapping in the park.*
Author.

protection and rehabilitation of the remaining lowland forests and wetlands, for
systematic research attempting to understand the ecology of these and other native
ecosystems, and for environmental education, is really much more important than
expenditure on, for example, general control of predators in the existing National
Parks. As Myers put it, with characteristic pithy wit, 'While conservation efforts need
to be greatly expanded, they need also to become more selective; the time is past
when we can achieve much by running hither and yon with buckets of water.'[15] In
New Zealand, we have hardly got past the stage of simple descriptive studies, whose
results can be interpreted in many ways. What we need are cast-iron field
experiments, and these need money and *time*. Good research and good planning,
says Alan Esler, a botanist who writes as if from bitter experience, cannot be done
with a bulldozer revving in the background.[16]

Epilogue

Can we make some final judgement about the past extinctions of so many native birds
from New Zealand? Can we say which of them could have been avoided, and which
could not, in a way that might help the future development of conservation policy? It
should be clear by now that the story is not so simple, and any such exercise would be
inviting an argument; all that can be given is mere opinion, to be accepted or not as
one chooses. Nevertheless, opinions can be interesting, and here is mine.

   I think that most of the historic extinctions were inevitable, given the biology of the

The relentless slaughter of native birds during the nineteenth century was sometimes justified on the grounds that museums needed specimens of species that were about to disappear anyway. That is true; but much of this irreplaceable material was sent overseas, not for safekeeping, but because overseas buyers could offer higher prices than could New Zealand's own adequate but young and poor institutions. As a result, both our forests and our national collections were impoverished for commercial rather than for scientific reasons. National Museum.

birds at risk and the changing conditions of their times. There is a long list of contributory causes which are often roundly and emotionally condemned by various writers and in various combinations, but I do not see that we could gain anything from censure at this stage, other than by compiling a list of Awful Examples to guide future decisions (for example, on the current argument about whether fishermen should be allowed to moor at the priceless rat-free Snares Islands (p.157)). Concerning the past extinctions only, I would not blame the birds themselves, since the attributes that accelerated their disappearance had previously served them very well for generation upon generation. I would not blame the Polynesian hunters or the early European explorers and surveyors, who encountered a fauna with no defences against their weapons, and who, driven by personal necessity, understandably used those weapons to good effect. I would not blame the cats or rats, or the ships that brought them, since their universal spread by the expanding European naval powers was unintentional and as unavoidable as death and taxes. I would not blame the mustelids, which arrived too late to add much extra damage, except in the south and west of the South Island. I would not blame the pioneer European farmers whose ignorance and desperation over-ruled the infant conservation movement of the 1880s, and whose lonely labours built the prosperous, civilized country New Zealand is now; and nor should anyone else who expects to enjoy a share of the relatively high standard of living that their descendants earn for us from the highly efficient farms that conservationists deplore. It is all very well to have noble regrets about the passing of the old order, but the effect is spoilt by hypocrisy.

Having said that, I think it is still possible to identify two groups of people who do deserve the censure of present and future generations. First, the collectors of the last century, who systematically hunted down the last of the rarest endemic birds for specimens far beyond the legitimate needs of science; second, all those who, in modern times, persist in causing or permitting the destruction of the last remaining forests and wetlands. In both cases the advantage was or is definitely on the human side of the equation, and was or is being exploited to the full – not for simple survival and in honest ignorance, but for calculated commercial profit,[e] regardless of consequences. Nothing that animal predators have done can equal what man is capable of doing for money: even rats, destructive though they are, have no conscious knowledge of or responsibility for their actions and, unlike man, cannot choose to behave in any other way. There is no doubt that the number one immigrant killer of modern times is man himself, especially the kind that comes equipped with chainsaws, bulldozers and mechanical diggers. The species man drives to extinction, unlike the tens of thousands of extinct species that strew the fossil record, are never replaced by new and different forms, only by the fuller expansion of the few that can live in man's world and on his terms.

All the same, there is a positive side to the changes man has precipitated on islands, one that could and should condition our attitude when we come to consider the future. On Lord Howe and the Hawaiian islands as well as in New Zealand, the extinction of some birds has been and is being counteracted by the recent natural colonization by others. These continual replacements emphasize that we must understand and allow for the fact that nature is *dynamic*. When landscapes, forests or the numbers of the animals that live in them appear to be unchanging, it is because

Animal populations are much more easily exterminated by habitat destruction than by direct persecution. Even Shakespeare knew that, when he put into Shylock's mouth the words 'you take my life when you do take the means whereby I live'. The most vulnerable endemic birds of New Zealand certainly have been wiped out by introduced animal predators in the past; but the hardy ones that are left are now threatened more often by various forms of habitat destruction than by predation. So which is the most significant immigrant killer of modern times? J. H. Johns, N.Z. Forest Service.

the conditions that govern them are varying relatively little from year to year. If the conditions change, permanently, the scene changes, irretrievably. This book is set out as a historical narrative in order to emphasize that conditions now are totally different from the way they were when man first arrived in New Zealand; and of course, dynamic nature has responded accordingly, impelled by biological forces completely beyond our control. It is absolutely irrational to think of trying to restore the remaining mainland forests to some former state, to 'preserve' them, as one does a priceless work of art, or to 'restore the balance of nature', because the endangered species have 'a right to exist'. There are a few places where things are still as they were in pre-European days; they are called museums. Conservation of nature in its best and widest sense is not a hanging on to the past, but a hanging on to the *future*; and no living thing has any 'right to exist', not even man.[f]

Evolution did not finish in 1840, and rational conservation policy must recognize the forces of change and the grander scale of things, including the situations where no management is possible or necessary. For example, the accelerated turnover of the New Zealand bird fauna in the human period is a natural process, even though man set it off; and, likewise, much though we may regret it, we have no choice but to accept stoats, red deer and possums as established members of our mainland forest communities. Changes wrought by man are only the latest that New Zealand has seen, and they will not be the last, even if we ourselves make no further new ones. The reverberations set off by what we have done already have not yet died away. R. K. Murton's definition of conservation, though written with reference to Britain, is particularly apt for New Zealand: 'Biological conservation is not synonymous with preservation or maintaining a *status quo*, but is instead concerned with sensible wildlife management in a dynamic environment constantly being altered by man.'[18]

The landscape of New Zealand has changed in a few centuries from mainly forest to mainly pastoral land; the sensitive endemic species are being replaced by hardier new arrivals from Australia, as well as man's own introductions from Europe and elsewhere, which are all accustomed to predators and to the now more abundant open-country habitats. Some native birds, like the harrier hawks, have greatly benefited from the changes and become more abundant than formerly, while others, which the early ornithologists expected to require special protection on island reserves (e.g. kea, bellbird, paradise duck)[19] are still common. We are not, after all, being left with no birds to watch – in fact, the total number of species in the list may even increase a little in the long run, because of the increased diversity of land habitats; and the abundant marine and coastal species have been affected by man's tampering with the landscape much less than have the inland birds. There is a danger of a certain maudlin sentimentality about the extinct and endangered endemic birds, which may cause us to forget that the new arrivals are often beautiful too, or are considered so in their own lands. Moreover, in evolutionary terms, the processes of their colonization and adaptation are quite as interesting as the processes of extinction. On the other hand, the birds we are losing, or have lost, are the unique ones, found nowhere else but in New Zealand, and the ones we are gaining are the common world tramps with little biological or (let it be said!) patriotic appeal. I have to admit that the exchange of sparrows and starlings, or even white-faced herons and spur-winged plovers, for laughing owls and huia, seems a bad bargain for us, and I am suggesting that we take a philosophical view of it only because it is too late to gain much by taking any other view.

In a nutshell, I think we should accept predators as permanent members of the New Zealand fauna; attempt to limit the damage they can do in the most sensitive remaining areas, but not waste money on impossible, general control of their populations; recognize that large-scale predator control on the mainland is not now necessary, even if it were possible, since the processes of nature are repopulating New Zealand with birds that are able to live with predators, while the rest are either adapting or have already gone; vigorously defend what isolated remnants of the ancient fauna are actually defensible in the long term; and temper our regret at the passing of the old endemics with positive appreciation of the new colonists.

NOTES AND REFERENCES

Notes and references are listed separately for each chapter, and do not overlap. The notes contain only information which amplifies the text, and their sources are listed with the chapter references, under superscript numbers integrated with those in the text. The chapter reference lists contain bibliographic details only, and each is independent of the others.

To minimize the number of entries I have referred to general sources at first mention only, and thereafter only to document quotations or especially important points; and I have mainly cited the most recent sources, which contain the references that will lead the serious reader to the earlier or related literature.

Notes to the Introduction

(a) In this context, and in similar contexts throughout this book, I use the word 'man' to mean 'mankind' (from the Latin, *Homo*) not 'masculine' (Latin *Vir*). No offence or sexist discrimination is intended, and none should be taken.

(b) Let no one infer a breath of ingratitude from this comment: it is merely a statement of fact that, had the investment in research on stoats in Fiordland during the 1970s been matched by a simultaneous and equal investment in research on the bird populations of the same areas, the conclusions with respect to birds (p.99) might have been more positive.[2]

References to the Introduction

1. J. A. Mills and G. R. Williams: The status of endangered New Zealand birds. In *The Status of Endangered Australasian Wildlife*, edited by M. J. Tyler. Royal Zoological Society of South Australia, Adelaide (1979), pp.147-168.
2. C. M. King: *Control of stoats in the National Parks and Reserves of New Zealand*. National Parks Scientific Series, Department of Lands & Survey, Wellington (in press).
3. T. Halliday: *Vanishing Birds*. Hutchinson, New Zealand (1978), p.19.
4. *Ibid*, p.20.
5. P. Whitfield: *The Hunters*. Hamlyn, London (1978).
6. C. M. King and R. L. Edgar: Techniques for trapping and tracking stoats (*Mustela eminea*): a review, and a new system. *NZ Journal of Zoology* 4: 193-212 (1977).

Notes to Chapter 1

(a) The area of the North Island is 114,453 sq km, and of the South Island 150,718 sq km. The largest of the other islands, Stewart, is 1,746 sq km, and the Chathams group totals 963 sq km. All the other islands, from the Kermadecs to the subantarctic Auckland and Campbell Islands, total 824 sq km. The grand total is 268,704 sq km, spread over an enormous area of ocean from 33°S to 53°S latitude and from 162°E to 173°W longitude.

(b) The explanation of the origin of the New Zealand rocks, fauna and flora presented here is the generally accepted one, as propounded in greater detail by C. A. Fleming[4]. However, there are alternatives.[5]

(c) The nearest land is Australia, 1900 km to the west; New Caledonia, 1500 km to the north; and South America, 9300 km to the east.

(d) An alternative suggestion is that island birds tend to become flightless in order to avoid being blown off the island by storms, but this is probably only one of the beneficial consequences of flightlessness, not its primary cause.

(e) There is still considerable argument as to what kind of habitat the undisturbed moa preferred, mainly because they and their habitats have been so much interfered with by man (p.50). Maybe at least some species also ate grass, but they certainly do not merit the popular description of 'lawn-moas'. It is true that forest was scarce, and grass abundant during the glacial stages; but these periods were short compared with the millions of years before the Ice Ages during which forest was the dominant habitat in New Zealand.

(f) In January 1770, Cook's crew anchored in Queen Charlotte Sound 'were awakened by the singing of the birds: the number was incredible, and they seemed to strain their throats in emulation of each other. This wild melody was infinitely superior to any that we had ever heard of the same kind; it seemed to be like small bells, most exquisitely tuned. . . .'[14]

(g) Theoretical ecologists are still arguing whether this gradation from opportunistic to equilibrium species, the so-called 'r-K spectrum', is of fundamental significance, or is just a convenient way of describing the observable differences in the life cycles of animals. I am using the terms just as labels. There are a lot of exceptions to the pattern, but it is still a useful concept.

References to Chapter 1

1. I. M. Wards (editor): *New Zealand Atlas*. Government Printer, Wellington (1976).
2. J. H. Johns and C. G. R. Chavasse: *The Forest World of New Zealand*. A. H. and A. W. Reed, Wellington (1975).
    B. Enting and L. Molloy: *The Ancient Islands*. Port Nicholson Press, Wellington (1982).
3. G. R. Stevens: *New Zealand Adrift*. A. H. and A. W. Reed, Wellington (1980).
4. C. A. Fleming: *The Geological History of New Zealand and its Life*. Auckland University Press and Oxford University Press, Auckland (1979).
5. R. C. Craw: Two biogeographical frameworks: implications for the biogeography of New Zealand. A review. *Tuatara* 23:81-114 (1978).
6. T. Halliday: *Vanishing Birds*. Hutchinson, New Zealand (1978), p.120.
7. C. A. Fleming, *op. cit.* p.84.
8. C. A. Fleming: History of the New Zealand land bird fauna. *Notornis* 9:270-274 (1962).
9. L. Cockayne: *New Zealand plants and their story* (4th edition, edited by E. J. Godley). N.Z. Government Printer, Wellington (1967).
10. G. Kuschel (editor): *Biogeography and ecology in New Zealand*. W. Junk Publishers, The Hague (1975).
    New Zealand's Nature Heritage, an encyclopaedia of N.Z. natural history published by Hamlyn in 105 parts between 1974 and 1976; pp.461-2, 257-62, and 29-32.
    G. R. Stevens, *op. cit.*, Chapter 14.
11. P. R. Millener: *The Quaternary avifauna of the North Island, New Zealand*. PhD Thesis, Auckland University (1981), p.765.
12. B. McCulloch: *No Moa. Some thoughts on the life and death of New Zealand's most spectacular bird*. Canterbury Museum, Christchurch (1982).
13. C. J. Burrows, B. McCulloch and M. M. Trotter: The diet of moas . . . *Records of the Canterbury Museum* 9(6):309-336 (1981).
14. Quoted in E. Best: *Forest Lore of the Maori*. Government Printer, Wellington (1942, reprinted 1977) p.112.
15. R. A. Falla, R. B. Sibson and E. G. Turbott: *The New Guide to the Birds of New Zealand*. Collins, Auckland (1981).
16. M. Gorman: *Island ecology*. Chapman & Hall, London (1979).
    M. Williamson: *Island Populations*. Oxford University Press, Oxford (1981).
17. G. Kuschel, *op cit.*, pp.248, 250.
18. R. H. McArthur and E. O. Wilson: *The Theory of Island Biogeography*. Princeton University Press, Princeton (1967).
19. J. A. Mills and G. R. Williams: The status of endangered New Zealand birds. In *The Status*

of Endangered Australasian Wildlife, edited by M. J. Tyler. Royal Zoological Society of South Australia (1979), p.147.

20. H. S. Horn: Optimal tactics of reproduction and life history. In *Behavioural Ecology: an evolutionary approach* (edited by J. R. Krebs and N. B. Davies). Blackwell Scientific Publications, Oxford (1978), pp.411-429.

 R. M. May (editor): *Theoretical Ecology: Principles and applications*, 2nd edition. Blackwell Scientific Publications, Oxford (1981), Ch.3.

21. A. Moorehead: *Darwin and the Beagle*. Hamish Hamilton, London (1969), p.161. The quotation is taken from p.386 of Darwin's own *Voyage of the Beagle* (1845).

22. M. H. Powlesland: A breeding study of the South Island Fantail (*Rhipidura f. fuliginosa*). *Notornis* 29: 181-95 (1982).

Notes to Chapter 2

(a) The precise date is unknown – the range is between 1200 and 1000 years ago.[1] Early theories about a double invasion by people of different races, retold by Lockley,[2] have no support from archaeologists.

(b) Perhaps mainly because they did not have to share the energy resources of the land with any mammals, which dominate animal communities elsewhere in the world.

(c) Mainly the southern elephant seal, the leopard seal, Hooker's sealion, and the New Zealand fur seal.[7]

(d) Archaeologists prefer the more accurate but unwieldy term 'the Archaic Phase of New Zealand Eastern Polynesian Culture' to 'moa-hunter', because not all the people of this period lived on moa – in parts of the country moa were seldom seen.[8] For the purposes of this book, however, it is the sections of that society that did hunt moa that we are most interested in.

(e) Some authors maintain that the arrival of the Polynesians also coincided with the end of a world-wide period of milder, damper climate, when the forests were at a short-term maximum, and already beginning to suffer from increased dryness. Kelly[12] speaks of them 'stepping ashore into a virtual tinderbox'. The evidence for and against this suggestion is reviewed briefly by McGlone,[11] who concludes that this effect, if any, was not very important.

(f) Actually, the hunters were not entirely deprived of large carcases for butchering. The fully developed 'classic' Maori culture was strongly war-like, and the frequent raids and battles were usually followed by cannibal feasts. However, the significance of cannibalism was not purely nutritional: to reduce one's defeated enemy to mere food was also a great psychological and spiritual victory.[15] As Cook remarked, 'they have a great liking for this kind of food'.[16]

 On the transition from 'Polynesian' to 'Maori', see p.60.

(g) This is no reflection on the needs of the Maoris or the sustaining properties of the New Zealand bush: it is a simple matter of energy supplies. In a mature forest, much of the primary productivity of the plant cover is locked up for years at a time in long-lived trees, whereas, when the forest is cleared, the plants that struggle to cover the bare land grow quickly and are shortlived, putting their energy into rapid annual growth or over-wintering stores (such as bracken root or potatoes) which are easily harvested by man. Throughout history, human activities have driven back forest and replaced it with the annual or biennial crops which can provide much more food per hectare than forest can. Deforestation has followed civilization from antiquity, a pattern recognized a hundred years ago by Marsh (p.63).

(h) This sentence was first written in 1981. Recently, bones of at least two extinct species of small birds, previously unknown to science, have been found in a cave in the Oparara Valley, Northwest Nelson (*Dominion*, 9 January 1984; *Forest and Bird*, November 1983).

References to Chapter 2

1. J. M. Davidson: The Polynesian Foundation. In *The Oxford History of New Zealand*, edited by W. H. Oliver and B. R. Williams. Oxford University Press, Wellington (1981), pp.3-27.

2. R. M. Lockley: *Man against Nature*. A. H. & A. W. Reed, Wellington (1970), Ch.2.

3. R. Duff: *The moa-hunter period of Maori culture*, 3rd edition. N.Z. Government Printer, Wellington (1977).

4. K. B. Cumberland: *Landmarks*. Reader's Digest, New South Wales (1981), p.48.

5. J. Davidson, *op. cit.*, p.6.

6. B. McCulloch: *No Moa. Some thoughts on the life and death of New Zealand's most spectacular bird*. Canterbury Museum, Christchurch, (1982).

7. A. Anderson: Faunal depletion and subsistence change in the early pre-history of southern New Zealand. *Archaeology in Oceania* 18: 1-10 (1983).

8. B. F. Leach: The pre-history of the southern Wairarapa. *Journal of the Royal Society of New Zealand* 11:11-33 (1981).

9. P. Houghton: *The first New Zealanders*. Hodder and Stoughton, Auckland (1980), Ch.6.
   A. Anderson: *When all the moa-ovens grew cold*. Otago Heritage Books, Dunedin (1983), p.8.

10. A. Anderson *ibid*, p.21.

11. M. S. McGlone: Polynesian deforestation of New Zealand: a preliminary synthesis. *Archaeology in Oceania* 18:11-25 (1983).

12. G. C. Kelly: Landscape and nature conservation. In *Land alone endures: Land use and the role of research* edited by L. F. Molloy *et al.* DSIR Discussion paper No. 4 (1980), pp.63-87.

13. C. J. Burrows and D. E. Greenland: An analysis of the evidence for climatic change in New Zealand in the last 1000 years. . . . *Journal of the Royal Society of N.Z.* 9:321-373 (1979).

14. J. Metge: *The Maoris of New Zealand* (revised edition). Routledge and Kegan Paul, London (1976), p.4.

15. *Ibid*, p.27.

16. A. H. and A. W. Reed (editors): *Captain Cook in New Zealand* (2nd edition). A. H. & A. W. Reed, Wellington (1969), p.206.

17. E. Best: *Forest Lore of the Maori*. Government Printer, Wellington (1977).

18. R. J. Cameron (1961) quoted in McGlone, *op. cit.*, p.23.

19. M. J. Daniel: Rats and Mice. In *New Zealand's Nature Heritage*. Paul Hamlyn Ltd, Hong Kong (1975), pp.2350-56.

20. E. Best, *op. cit.*, p.361.

21. J. Meeson: The plague of rats in Nelson & Marlborough. *Transactions and Proceedings of the N.Z. Institute* 17:199-207 (1885).
    See also J. Rutland, *loc. cit.* vol 22:300-307; and F. W. Hutton, *loc. cit.* vol 20:43. For an alternative view, see E. Best, *op. cit.*, p.354.

22. G. M. Thomson: *The naturalization of plants and animals in New Zealand*. Cambridge University Press, Cambridge (1922) p.64.

23. A. H. and A. W. Reed, *op. cit.*, p.178.

24. P. R. Millener: *The Quaternary avifauna of the North Island of New Zealand*. PhD Thesis, Auckland University (1981), p.761, and unpublished.
    P. R. Millener and C. J. Templer: The subfossil deposits of Paryphanta (Mac's Quarry) Cave, Waitomo. *Journal of the Royal Society of N.Z.* 11:157-166 (1981).
    C. A. Fleming: The extinction of moas and other animals during the Holocene period. *Notornis* 10:113-117 (1962).

25. R. J. Scarlett: Moa and man in New Zealand. *Notornis* 21:1-12 (1974). H. M. Leach, in *New Zealand's Nature Heritage* (1974), p.122.

26. P. S. Martin: Prehistoric Overkill. In *Pleistocene extinctions: the search for a cause*, edited by P.

S. Martin and H. E. Wright. Yale University Press, New Haven (1967) pp.75-120, and M. Coe: The bigger they are.... *Oryx* 16: 225-228 (1982). However, many people disagree with this view, and data on North America birds do not support it – see D. K. Grayson: Pleistocene avifaunas and the overkill hypothesis. *Science* 195: 691-693 (1977).

27. W. R. B. Oliver: The moas of New Zealand and Australia. *Bulletin of the Dominion Museum* 15:1-206 (1949).

28. T. Kirk: The displacement of species in New Zealand. *Transactions and Proceedings of the N.Z. Institute* 28:1-27 (1896).

29. S. J. Gould: *Ever since Darwin*. W. W. Norton, New York (1977), p.84.

30. S. J. Gould: *Hens' Teeth and Horses' Toes*. W. W. Norton, New York (1983), pp.343-352.

31. G. R. Williams: Extinction and the land and freshwater inhabiting birds of New Zealand. *Notornis* 10:15-32 (1962).

32. W. Dritschilo, H. Cornell, D. Nafus and B. O'Connor: Insular biogeography: of mice and mites. *Science* 190:467-469 (1975).

33. A. Bathgate: Some changes in the fauna and flora of Otago in the last sixty years. *N.Z. Journal of Science & Technology* 4:273-283 (1922).

34. P. R. Millener, *op. cit.*, p.767-8.

35. A. Anderson (reference 7 above), p.6.

36. R. J. Scarlett, *op. cit.*, p.7.

37. K. B. Cumberland, *op. cit.*, p.44.

38. A. Anderson (reference 9 above), p.24.

39. R. A. Falla: New Zealand bird life, past and present. *Cawthron Lecture Series* 29:1-15 (1955).

40. M. S. McGlone, *op. cit.*, p.23.

41. E. Best, *op. cit.*, p.186-7.

42. I. A. E. Atkinson: Evidence for the effects of rodents on the vertebrate wildlife of New Zealand islands. In *The ecology and control of rodents in New Zealand nature reserves*, edited by P. R. Dingwall, I. A. E. Atkinson and C. Hay. Department of Lands & Survey, Information series 4:7-31 (1978).

43. A. H. Whitaker: Lizard populations on islands with and without Polynesian rats, *Rattus exulans* (Peale). *Proceedings of the N.Z. Ecological Society* 20:121-130 (1973).
    I. G. Crook: The tuatara, *Sphenodon punctatus* (Gray), on islands with and without populations of the Polynesian rat, *Rattus exulans* (Peale). *Proceedings of the N.Z. Ecological Society* 20:115-120 (1973).

44. J. A. D. Flack and B. D. Lloyd: The effect of rodents on the breeding success of the South Island Robin. In *The ecology & control of rodents in New Zealand nature reserves*, edited by P. R. Dingwall, I. A. E. Atkinson, and C. Hay. Department of Lands & Survey Information Series 4:59-66 (1978).

45. P. V. Kirch: Man's role in modifying tropical and subtropical Polynesian ecosystems. *Archaeology in Oceania* 18:26-31 (1983).

Notes to Chapter 3

(a) The 'land wars' were not only about land, but also about the power, authority and *mana* (prestige) of the leaders of the two races over the land and the people it sustained.[18]

(b) The post-war land confiscations were decided on according to land value, not to the degree of punishment deserved. The Waikato tribes, who owned large areas of what is now the best farmland in the country; lost almost everything – and included among them were some 'friendly' tribes who had actually supported the government during the war. By contrast, some of the more aggressive hill-country tribes in the King Country got off lightly. The original confiscations added up to over 3 million acres (1.2 million hectares) but some were later paid for or returned.[23] The King Country forests remained undisturbed for many more years, and are now among the last remaining strongholds of the North Island kokako (p.146).

(c) This account is entered by Thomson (1922, p.79) under *R. rattus*, but if Atkinson is right, they were more likely to have been *R. norvegicus*.

(d) *Mus decumanus* is the old name for *Rattus norvegicus*.

(e) From what Gillies says later in this passage, the second part of this extract could have been referring to kiore, but the first part is certainly about *norvegicus*. Both parts give a powerful impression of how much more abundant rats were in those days than now.

(f) Perhaps this was a short-term irruption, rather than the usual population density of rats in Fiordland. At any rate, ground birds were still common ten years later, when Richard Henry was working in Fiordland; then, around the turn of the century, both ground birds and rats became scarce.

(g) The Australian stubble quail still survives in its homeland, despite all the introduced predators and other disturbances that assail it there; and in the 1860s and 1870s it was also introduced here, but without success (Thomson, 1922, p.3).

References to Chapter 3

1. *Abel Tasman National Park Handbook*. Department of Lands & Survey, Wellington (1976).
2. J. M. R. Owens: New Zealand before Annexation. In *The Oxford History of New Zealand*, edited by W. H. Oliver and B. R. Williams. Oxford University Press, Wellington (1981), pp.28-53.
3. A. C. Begg and N. C. Begg: *Dusky Bay*. Whitcombe & Tombs, Christchurch (1966).
4. J. Owens, *op. cit.*, p.32.
5. A. Voice: The subantarctic Islands. Towards Preservation in perpetuity. *Landscape* 6: 5-8 (1979).
6. R. McNab: *The old whaling days* (1913). Reprinted by Golden Press Pty, Auckland (1975), p.82.
7. J. Owens, *op. cit.*, p.33.
8. J. Owens, *op. cit.*, p.44; G. Ell: *Wild Islands*. Bush Press, Auckland (1982).
9. J. Owens, *op. cit.*, p.35.
10. E. Dieffenbach: *Travels in New Zealand*, vol 1. J. Murray, London (1843), pp.227-28.
11. W. J. Gardner: A colonial economy. In *The Oxford History of New Zealand*, edited by W. H. Oliver and B. R. Williams. Oxford University Press, Wellington (1981), pp.57-86.
12. K. B. Cumberland: *Landmarks*. Reader's Digest, Sydney (1981), p.84.
13. J. Owens, *op. cit.*, p.39.
14. J. Owens, *op. cit.*, p.49.
15. J. Metge: *The Maoris of New Zealand*, revised edition. Routledge and Kegan Paul, London (1976), p.76.
16. *Ibid.*, p.31.
17. M. P. K. Sorrenson: Maori and Pakeha. In *The Oxford History of New Zealand*, edited by W. H. Oliver and B. R. Williams. Oxford University Press, Wellington (1981), pp.168-193.
18. *Ibid*, p.175.
19. W. J. Gardner, *op. cit.*, p.62, 64.
20. K. B. Cumberland, *op. cit.*, p.109.
21. J. Hall-Jones: *Mr Surveyor Thomson*. A. H. & A. W. Reed, Wellington (1971), pp.59,60,67-68.
22. J. T. Holloway: *The Mountain Lands of New Zealand*. Tussock Grasslands & Mountain Lands Institute, Lincoln (1982), p.13.
23. M. Sorrenson, *op. cit.*, p.186.

24. W. Gardner, *op. cit.*, p.73.

25. G. P. Marsh: *Man and Nature* (1864). Reprinted by the Belknap Press of Harvard University Press, Cambridge, Mass. (1965).

26. D. J. Wishart: *The fur trade of the American West 1807-1840*. Croom Helm, London (1979), p.207.

27. G. M. Thomson: *The naturalization of animals & plants in New Zealand*. Cambridge University Press, Cambridge (1922), p.22.

28. A. H. Clarke: *The invasion of New Zealand by people, plants & animals. The South Island*. Rutgers University Press, New Brunswick (1949), p.266.

29. F. C. Kinsky (editor). *Annotated checklist of the birds of New Zealand* (1970), with amendments and additions as listed in *Notornis* 27 (1980). Ornithological Society of New Zealand Inc. A. H. & A. W. Reed, Wellington.

   K. Wodzicki: The status of some exotic vertebrates in the ecology of New Zealand. In *The genetics of colonising species*, edited by H. G. Baker and G. L. Stebbins. Academic Press, New York (1965), pp.425-60.

   J. A. Gibb and J. E. C. Flux: Mammals. In *The natural history of New Zealand*, edited by G. R. Williams. A. H. & A. W. Reed, Wellington (1973), pp.334-371.

30. J. P. Skipworth: Aliens in New Zealand. *N.Z. Journal of Ecology* 6:145-6(1983).

   J. Druett: *Exotic Intruders*. Heinemann Publishers, Auckland (1983).

31. Gibb & Flux, *op.cit.*, p.343.

32. C. Hursthouse: *New Zealand, or Zealandia, the Britain of the South*. Stanford, London (1851), p.117.

33. I. A. E. Atkinson: Spread of the ship rat (*Rattus r. rattus* L.) in New Zealand. *Journal of the Royal Society of N.Z.* 3:457-472(1973).

34. A. H. and A. W. Reed: *Captain Cook in New Zealand*. A. H. & A. W. Reed, Wellington (1969), p.50.

35. F. von Hochstetter: *New Zealand: its physical geography, geology and natural history*. Stuttgart (1867).

36. G. M. Thomson, *op.cit.*, p.77.

37. I. A. E. Atkinson: Evidence for the effects of rodents on the vertebrate wildlife of New Zealand islands. In *The ecology & control of rodents in New Zealand nature reserves*, edited by P. R. Dingwall, I. A. E. Atkinson and C. Hay. Department of Lands & Survey, Information Series 4:7-30(1978).

38. G. M. Thomson, *op.cit.*, p.68.

39. Quoted in J. C. Greenway: *Extinct and vanishing birds of the world*. American Committee for International Wildlife protection, New York (1958), p.63.

40. A. H. and A. W. Reed, *op.cit.*, p.159.

41. R. McNab, *op.cit.*, p.349.

42. I. A. E. Atkinson, *op.cit.*, (1973) and in press (ICBP Technical Publication, Cambridge).

43. W. L. Buller: Further notes on the ornithology of New Zealand. *Transactions of the N.Z. Institute* 3:37-56(1871).

44. G. M. Thomson, *op.cit.*, p.82.

45. R. Gillies: Notes on some changes in the fauna of Otago. *Transactions of the N.Z. Institute* 10:306-322(1878).

46. A. Reischek: *Yesterdays in Maoriland*, translated by H. E. L. Priday. Jonathan Cape, London (1930), p.251.

47. Quoted in Atkinson (1973), p.465.

48. J. G. Innes and J. P. Skipworth: Home ranges of ship rats in a small New Zealand forest as revealed by trapping and tracking. *N.Z. Journal of Zoology* 10:99-110(1983).

49. B. D. Bell: The Big South Cape Islands rat irruption. In *The ecology and control of rodents in*

New Zealand nature reserves, edited by P. R. Dingwall, I. A. E. Atkinson and B. D. Bell. Department of Lands & Survey, Information Series 4:33-37(1978).

50. M. J. Daniel: Seasonal diet of the ship rat. (*Rattus r. rattus*) in a lowland forest in New Zealand. *Proceedings of the N.Z. Ecological Society* 20:21-30(1973).

51. Quoted by Begg and Begg, *op.cit.*, p.170.

52. Thomson, *op.cit.*, p.61.

53. I. A. E. Atkinson and B. D. Bell: Offshore and outlying islands. In *The Natural History of New Zealand*, edited by G. R. Williams, A. H. & A. W. Reed, Wellington (1973), pp.372-392.

54. D. V. Merton: Controlling introduced predators and competitors on islands. In *Endangered birds*, edited by S. A. Temple. Croom Helm, London (1978), pp.121-128.

55. Thomson, *op.cit.*, p.64.

56. J. Pascoe (ed): *Mr Explorer Douglas*. A. H. & A. W. Reed, Wellington (1957), p.275.

57. A. P. Harper: *Pioneer work in the Alps of New Zealand*, T. Fisher Unwin, London (1896), p.43.

58. B. J. Karl and H. A. Best: Feral cats on Stewart Island; their foods, and their effect on kakapo. *N.Z. Journal of Zoology* 9:287-294(1982).

59. B. M. Fitzgerald and B. J. Karl: Foods of feral housecats (*Felis catus* L.) in forest of the Orongorongo Valley, Wellington. *N.Z. Journal of Zoology* 6:107-126(1979).
    B. M. Fitzgerald, W. B. Johnson, C. M. King and P. J. Moors: *A review of research on mustelids and cats in New Zealand*. Wildlife Research Liaison Group, Department of Internal Affairs, Wellington (1984).

60. P. J. Moors: Methods for studying predators and their effects on forest birds. In *The ecology and control of rodents in New Zealand nature reserves*, edited by P. R. Dingwall, I. A. E. Atkinson and C. Hay. Department of Lands & Survey, Information Series 4:47-57(1978).

61. R. H. Taylor: How the Macquarie Island parakeet became extinct. *N.Z. Journal of Ecology* 2:42-45(1979).

62. G. M. Thomson, *op.cit.*, p.35.

63. *Ibid*, p.38.

64. E. Best: *Forest Lore of the Maori*, Government Printer, Wellington (1977), p.182.

65. R. Owen, in *Proceedings of the Zoological Society of London* 120:8-10(1843).

66. A. H. and A. W. Reed, *op.cit.*, pp.51, 161, 166, 176, 177.

67. J. Pascoe, *op.cit.*, p.61.

68. E. J. Wakefield: *Adventure in New Zealand*. J. Murray, London (1845), pp.56, 58.

69. M. Tracy: *West Coast Yesterdays*, A. H. & A. W. Reed, Wellington (1960), p.174.

70. J. D. Enys: An account of the Maori manner of preserving the skin of the Huia, *Heteralocha auctirostris*, Buller. *Transactions of the N.Z. Institute* 8:204-205(1876).

71. T. H. Potts: On the birds of New Zealand. *Transactions of the N.Z. Institute* 2:40-78(1870).

72. M. King: *The Collector*. Hodder & Stoughton, Auckland (1981), pp.66, 109-10.

73. W. L. Buller: Notes on New Zealand Ornithology. *Transactions and Proceedings of the N.Z. Institute* 28:326-358(1896).

74. M. King, *op.cit.*, p.40.

75. N. Myers: *The Sinking Ark*. Permagon Press, Oxford (1979), p.94-5.

76. A. Bathgate: Some changes in the fauna and flora of Otago in the last sixty years. *N.Z. Journal of Science and Technology* 4:273-283(1922).

77. W. L. Buller: On the disappearance of the korimako (*Anthornis melanura*) from the North Island. *Transactions of the N.Z. Institute* 10:209-211(1878).

78. M. Harrison: The orange-fronted parakeet, *Cyanoramphus malherbi*. *Notornis* 17:115-125(1970).

F. N. Hayes and M. Williams: The status, aviculture and re-establishment of brown teal in New Zealand. *Wildfowl* 33:73-80(1982).

79. E. Best, *op.cit.,* p.179.
80. W. J. Phillips: *The Book of the Huia.* Whitcombe & Tombs Ltd, Christchurch (1963), p.60.
81. G. R. Williams: Birds. In *The Natural History of New Zealand*, edited by G. R. Williams. A. H. & A. W. Reed, Wellington (1973), p.314.
82. E. Best, *op.cit.,* p.116.

Notes to Chapter 4

(a) The individual Acclimatization Societies acted independently, and even in opposition to each other; some helped to bring in the mustelids, others vigorously protested.

(b) The reasons for Martin's perfectly correct observation can be explained by modern ecological theory.[5]

(c) All energy comes from the sun, and is passed along the food chain from plants to herbivores and then to one or more carnivores. At every step in the chain, most of the energy is used or lost, and very little is stored and passed on to the animals waiting at the next step. For an elegant and simple explanation of this and all the other basic principles of ecology, see *Why Big Fierce Animals are Rare* by Paul Colinvaux.[39]

(d) The work summarized in these few pages has been published in great detail elsewhere: see Fitzgerald *et al*[34] for full bibliographies of New Zealand work on ferrets, weasels and stoats.

(e) Early accounts often mention 'weasels' when it seems from what we know today that the animal was more likely to have been a stoat. We cannot tell now whether weasels (*Mustela nivalis*) really were more common in forests than stoats then (although the opposite is the case now), or whether the early writers were following the custom, still usual in North America now, of using the word 'weasel' in a more general sense, to include all the small members of the weasel family, the Mustelidae.

(f) An alternative possible explanation is that stoats thrived better in forests where they could avoid inteference competition from ferrets and/or predation from harriers. Note that Okuru is only 45 map-kilometres from Makaroa (see below). Stoats liberated there could easily have reached Okuru in a few months; in January 1980 in the Hollyford Valley, a young male stoat, marked with eartags, travelled at least 24 km in seven days.[41]

(g) The Makaroa Valley was one of the earliest release sites (p.87).

(h) However, the memories of old-timers, no doubt correct in outline, are often astray in the details that count. For example, a well-known Te Anau entity stated in the *Southland Times* of 17 November 1967 that '60 years ago the Hollyford was like a fowlhouse'; yet, according to contemporary reports in the *Otago Witness*, the weka, kiwi and kakapo were almost exterminated in the Hollyford by 1890 (p.101).

(i) New Zealanders might be surprised to hear that their country has been used overseas as an example of poor conservation and over-development.[82] In fact, contemporary New Zealand farmlands bear an uncomfortable resemblance to the prophetic description of John Stuart Mill: 'nothing left to the spontaneity of nature . . . all quadrupeds or birds which are not domesticated for man's use exterminated as his rivals for food . . . scarcely a place left where a wild shrub or flower [read "native"] could grow without being eradicated in the name of improved agriculture' (J. S. Mill, 1848, quoted with illuminating comment by Passmore[83]).

References to Chapter 4

1. G. M. Thomson: *The Naturalization of animals & plants in New Zealand.* Cambridge University Press, Cambridge (1922), p.87.

2. L. W. McCaskill: *Hold this land*, A. H. & A. W. Reed, Wellington (1973), p.167.

3. W. L. Buller: On the proposed introduction of the Polecat into New Zealand. *Transaction and Proceedings of the N.Z. Institute* 9:634-5(1877).

4. G. R. Potts: The effects of modern agriculture, nest predation and game management on the population ecology of partridges . . . *Advances in Ecological Research* 11:1-79(1980).

5. C. M. King and P. J. Moors: The life history tactics of mustelids, and their significance for predator control and conservation in New Zealand. *N.Z. Journal of Zoology* 6:619-22(1979).

6. H. B. Martin: Objections to the introduction of beasts of prey to destroy the rabbit. *Transactions and Proceedings of the N.Z. Institute* 17:179-182(1885).

7. A. Reischek: Observations on the habits of New Zealand birds, their usefulness or destructiveness to the country. *Transactions and Proccedings of the N.Z. Institute* 18:96-104(1886).

8. J. E. Harting: The weasel. *Zoologist* 52 (1894), pp.417-23 and 445-54.

9. W. H. McLean: *Rabbits Galore!* A. H. & A. W. Reed, Wellington (1966), pp.94, 106.

10. P. Walsh: The effect of deer on the New Zealand bush; a plea for the protection of our forest reserves. *Transactions and Proceedings of the N.Z. Institute* 25:435-439(1893).

11. W. L. Buller: Some curiosities of birdlife. *Transactions and Proceedings of the N.Z. Institute* 27:134-142(1895).

12. R. Chisholm: The work of acclimatization. *Otago Witness*, 18 December 1907.

13. H. Guthrie-Smith: *Tutira* (4th edition). A. H. & A. W. Reed, Wellington (1969), pp.349, 354.

14. G. M. Thomson, *op.cit.*, p.73.

15. *Ibid*, p.70.

16. K. Wodzicki: *The introduced mammals of New Zealand*. DSIR Bulletin 98 (1950) pp.75, 79-80.

17. Quoted by P. and A. Ehrlich: *Extinction: The causes and consequences of the disappearance of species*. Victor Gollancz, London (1982), p.233.

18. W. L. Buller: Notes on New Zealand Ornithology, with an exhibition of specimens. *Transactions and Proceedings of the N.Z. Institute* 28:326-358(1896).

19. E. Melland: Notes on a paper entitled 'The Takahe in western Otago' by Mr James Park, FGS. *Transactions and Proceedings of the N.Z. Institute* 22:295-300(1890).

20. T. Brooking: Economic Transformation. In *The Oxford History of New Zealand*, edited by W. H. Oliver & B. R. Williams. Oxford University Press, Wellington (1981), pp.226-249.

21. J. Metge: *The Maoris of New Zealand*, revised edition. Routledge & Kegan Paul, London (1976), p.12.

22. *Ibid*, p.109.

23. M. P. K. Sorrenson: Maori and Pakeha. In *The Oxford History of New Zealand*, edited by W. H. Oliver & B. R. Williams. Oxford University Press, Wellington (1981), pp.168-193.

24. L. Cobb and J. Duncan: *New Zealand's National Parks*. P. Hamlyn, Auckland (1980), p.12.

25. J. Metge, *op.cit.*, p.37.

26. M. Sorrenson, *op.cit.*, p.193.

27. W. A. Pullar: Periods of recent infilling of the Gisborne Plains basin: associated marker beds and recent changes in the shoreline. *N.Z. Journal of Science* 13:410-434(1970).
    K. B. Cumberland: *Landmarks*. Reader's Digest, Sydney (1981), Chapter 7.

28. L. F. Molloy *et al*: *Land Alone Endures*. DSIR Discussion Paper 3: (1980), p.12.

29. R. B. Brockie: Road mortality of the hedgehog (*Erinaceus europaeus* L.) in New Zealand. *Proceedings of the Zoological Society of London* 134:505-508(1960).

30. R. B. Brockie: Observations on the food of the hedgehog (*Erinaceus europaeus* L.) in New Zealand. *N.Z. Journal of Science* 2:121-136(1959).

31. R. J. Roser and R. B. Lavers: Food habits of the ferret (*Mustela putorius furo* L.) at Puke Puke Lagoon, New Zealand. *N.Z. Journal of Zoology* 3:269-75(1976).

32. W. H. Marshall: The ecology of mustelids in New Zealand. *DSIR Information Series* 38:1-32(1963)

33. C. M. King and J. E. Moody: The biology of the stoat (*Mustela erminea*) in the National Parks of New Zealand. *N.Z. Journal of Zoology* 9:49-144(1982).

34. C. M. King: The relationships between beech (*Nothofagus* sp.) seedfall and populations of mice (*Mus musculus*), and the demographic and dietary responses of stoats (*Mustela erminea*), in three New Zealand forests. *Journal of Animal Ecology* 52:141-166(1983).
   B. M. Fitzgerald, W. B. Johnson, C. M. King and P. J. Moors: *A review of research on mustelids and cats in New Zealand.* Wildlife Research Liaison Group, Department of Internal Affairs, Wellington (1984).

35. C. M. King and P. J. Moors: On coexistence, foraging strategy and the biogeography of weasels and stoats (*Mustela nivalis* and *M. erminea*) in Britain. *Oecologia* 39:129-150(1979).

36. C. M. King: Population biology of the weasel *Mustela nivalis* on British game estates. *Holarctic Ecology* 3:160-168(1980).

37. C. M. King and J. E. Moody, *op.cit.,* p.71.

38. E. Dunn: Predation by weasels (*Mustela nivalis*) on breeding tits (*Parus* sp) in relation to the density of tits and rodents. *Journal of Animal Ecology* 46:633-652(1977).

39. P. Colinvaux: *Why big fierce animals are rare.* Princeton University Press, Princeton (1978).

40. A. P. Harper: *Pioneer work in the Alps of New Zealand.* T. Fisher Unwin, London (1896), p.225.

41. C. M. King and C. D. MacMillan: Population structure and dispersal of peak-year cohorts of stoats (*Mustela erminea*) in two New Zealand forests, with special reference to control. *N.Z. Journal of Ecology* 5:59-66(1982).

42. Quoted in P. Houghton: *Hidden Water.* Hodder & Stoughton, Auckland (1974), p.120.

43. G. M. Thomson, *op.cit.,* p.74.

44. J. Pascoe (ed): *Mr Explorer Douglas.* A. H. & A. W. Reed, Wellington (1957), p.264.

45. W. L. Buller: Further notes and observations on certain species of New Zealand birds (with exhibits). *Transactions and Proceedings of the N.Z. Institute* 24:75-91(1892).

46. J. Pascoe, *op.cit.,* p.259-260.

47. *Ibid,* p.174.

48. A. P. Harper, *op.cit.,* p.291.

49. *Ibid,* p.149-50.

50. W. Buller (1895) *op.cit.,* p.138-139.

51. A. C. and N. C. Begg: *Dusky Bay.* Whitcombe & Tombs Ltd, Christchurch (1966), pp.164, 172.

52. A. C. and N. C. Begg: *Port Preservation.* Whitcombe & Tombs Ltd, Christchurch (1973), p.307.

53. P. Houghton, *op.cit.,* pp.119, 122.

54. C. M. King: *Control of stoats in the National Parks and Reserves of New Zealand.* National Parks Scientific Series, Department of Lands & Survey, Wellington (in press).

55. J. A. Gibb and J. E. C. Flux: Mammals. In *The Natural History of New Zealand*, edited by G. R. Williams. A. H. & A. W. Reed, Wellington (1973), pp.334-371.

56. M. J. Daniel: Population ecology of ship and Norway rats in New Zealand. In *The ecology and control of rodents in New Zealand nature reserves*, edited by P. R. Dingwall, I. A. E. Atkinson and C. Hay. Department of Lands & Survey, Information Series 4 (1978), pp.145-152.

57. J. S. Watson: The present distribution of *Rattus exulans* (Peale) in New Zealand. *N.Z. Journal of Science & Technology* 37:560-570(1956).

58. M. E. Soulé and B. A. Wilcox: *Conservation biology*. Sinauer Associates, Mass. (1980), p.105.

59. D. G. Dawson: Principles of ecological biogeography and criteria for reserve design. *Journal of the Royal Society of N.Z.* 14:11-15 (1984).

60. A. Saunders: *Wildlife and lowland forests*. N.Z. Wildlife Service, Fauna Survey Unit Report No. 27 (1980).

61. K. R. Hackwell: The Island Biogeography of native forest birds and reserve design. In *Proceedings of a workshop on a biogeographic framework for planning an extended National Parks and Reserves system*. Department of Lands & Survey, Wellington, Reserves Series 10:28-40(1982).

62. G. Ell: *Wild Islands*. Bush Press, Auckland (1982).

63. A. H. & A. W. Reed: *Captain Cook in New Zealand*. A. H. & A. W. Reed, Wellington (1969), p.176.

64. P. Walsh, *op.cit.*, pp.435-6, 437, 438.

65. C. M. H. Clarke: Eruption, deterioration and decline of the Nelson red deer herd. *N.Z. Journal of Forestry Science* 5:235-249(1976).
    G. Caughley: *The deer wars*. Heinemann, Auckland (1983), pp.4-6.

66. Gibb and Flux, *op.cit.*, p.357.

67. A. F. Mark and G. T. S. Baylis: Further studies on the impact of deer on Secretary Island, Fiordland, New Zealand. *N.Z. Journal of Ecology* 5:67-75(1982).

68. M. J. Mead: Effects of opossum browsing on northern rata trees in the Orongorongo Valley, New Zealand. *N.Z. Journal of Zoology* 3:127-139(1976).

69. A. E. Fitzgerald and P. Wardle: Food of the opossum *Trichosurus vulpecula* (Kerr) in the Waiho Valley, South Westland. *N.Z. Journal of Zoology* 6:339-345(1979).

70. P. Child: *Birdlife of Mount Aspiring National Park*. National Parks Scientific Series 4:1-64(1981).

71. G. Caughley, *op.cit.*, p.71.
    T. T. Veblen and G. H. Stewart: The effects of introduced wild animals on New Zealand forests. *Annals of the Association of American Geographers* 72:372-397(1982).

72. D. V. Merton: Controlling introduced predators and competitors on islands. In *Endangered Birds*, edited by S. A. Temple. Croom Helm, London (1978), pp.121-128.

73. M. R. Rudge and T. J. Smit: Expected rate of increase of hunted populations of feral goats (*Capra hircus* L.). *N.Z. Journal of Science* 13:256-259(1970).

74. G. M. Thomson, *op.cit.*, p.509.

75. J. M. Diamond and C. R. Veitch: Extinctions and introductions in the New Zealand avifauna: cause and effect? *Science* 211:499-501(1981).

76. H. Guthrie-Smith, *op.cit.*, p.202-3.

77. C. L. McLay: The species diversity of New Zealand forest birds: some possible consequences of the modification of beech forests. *N.Z. Journal of Zoology* 1:179-196(1974).

78. A. S. Garrick: Diets of pipits and skylarks at Huiarua Station, Tokomaru Bay, North Island, New Zealand. *N.Z. Journal of Ecology* 4:106-114(1981).

79. G. M. Thomson, *op.cit.*, p.510.

80. J. A. Mills and G. R. Williams: The status of endangered New Zealand birds. In *The Status of Australasian Wildlife*, edited by M. J. Tyler. Royal Zoological Society of South Australia (1979), p.153.

81. R. O'Regan: Effects of stoats on native birds. *Forest and Bird* 162 (1966), pp.5-7.

82. K. Garratt: A philosophy for nature conservation in New Zealand. *Landscape* 7:21-23(1980).

83. J. Passmore: *Man's responsibility for nature* (2nd edition). Duckworth, London (1980), p.121.

Notes to Chapter 5

(a) 'Introduced predators' in this context does not only mean the aliens brought from overseas. The native wekas introduced from the mainland onto offshore islands from which they were naturally absent, e.g. Macquarie (p.75), Codfish and Open Bay Islands, have been almost as bad.[1]

(b) The Royal Forest and Bird Society is a responsible body, which had not jumped to the wrong conclusion concerning the purpose and techniques of this study, as had the authors of these ignorant criticisms. But to save the Fiordland National Park Board any unnecessary embarrassments, the *Southland Times* printed an article explaining the misunderstanding on 27 November 1980.

(c) Stoats are carnivores, not rodents.

(d) The 'fool', the 'ass' and the 'brainless one' were all the same person, myself. In addition to this crime, I have been responsible for several other research projects on stoats in previous years, going back to 1971, as summarized in my final report.[2] Anyone interested in explanations of why those studies were done can find them there, and here; but I expect that those who phrase their comments in such terms as these would rather avoid having to be concerned with facts.

(e) The biomass of birds on Kapiti is reckoned by R. E. Brockie and A. Moeed[21] to be four to ten times greater than in the Orongorongo Valley, west of Wellington. They attribute the difference to predation on the mainland. Saddlebacks and stitchbirds are now being transferred to Kapiti from the Hauraki Gulf, though so far the saddlebacks are not doing very well, at least partly because of the Norway rats.[22]

(f) Black robins have the same characteristics (long lifespan and ability to replace lost clutches) as South Island robins have. The Wildlife Service is therefore able to apply intensive management techniques such as cross-fostering, which involves transferring the robin's first clutch of eggs to be reared by fosterparents (at first, Chatham Island warblers, and later, Chatham Island tits), leaving the robin parents to lay again and rear a second brood themselves. This technique helps the robins to achieve nearly double the annual productivity that they could manage alone. When this programme started in spring (September) 1980, the world population of robins was only five; by the end of the 1981/82 season, after only two seasons, it had more than doubled to 12: in March 1984, it reached 20.[17]

(g) New Zealanders accustomed to being on the receiving end of advanced American technology may be pleased to hear that some Kiwi farming expertise has made the return trip. Electric fences used to protect ducks nesting on the open prairies of the American west were powered by a new system of 12-volt energizers made in New Zealand. The nesting success of the ducks inside the fence was doubled.[34]

(h) The hypothesis that takahe could be at particularly increased risk from stoats after a beech seedfall, though put forward by J. A. Mills at the Takahe Seminar in 1978 only as a suggestion, has now achieved the status of a full and satisfactory explanation – appearing, for example, without any qualification or reference in several recent general natural history books.

(i) There are no offshore islands with suitable habitat, and years of attempts to breed takahe in captivity (as distinct from raising young from eggs taken from the wild) have been constantly dogged by infertility and plain bad luck. In the 11 years 1960-71, 44 eggs were laid at Mount Bruce, all infertile. From 1976-79, eight chicks were hatched, of which four survived;[46] in 1983, the first second-generation chick was produced by a female takahe reared in 1978, but it was later killed, probably by a stoat (*Evening Post*, 17.2.84).

(j) The first stoat was seen, but not caught, in May 1982. In the following summer, two
    juvenile males were trapped on the island; by spring (September) 1983, five females and
    another male had been caught. The ovaries of the four females that could be examined
    were without corpora lutea,[55] which means that there was no adult male present on the
    island in summer 1982-83; the eight caught could all be members of one litter.[56] This
    suggests that the single stoat seen in May 1982 was a female. All female stoats are pregnant
    all the time, with eight to ten potential embryos in a state of suspended animation (p.98).
    If the intensive trapping campaign mounted by the Wildlife Service has left any females
    on the island, the earliest they could conceive would be September 1983, and the earliest
    they could produce a litter would be September 1984. By then they might have died,
    swum away, or been caught in traps, so there is hope of clearing stoats off Maud Island in
    the near future. But what has happened once could presumably happen again, and the
    suitability of Maud as a long-term refuge for endangered species must now be
    permanently compromised, despite official talk about 'facing up to the challenge'.[57]

References to Chapter 5

1. I. A. E. Atkinson and B. D. Bell: Offshore and outlying islands. In *The natural history of New
    Zealand*, edited by G. R. Williams. A. H. & A. W. Reed, Wellington (1973), pp.381, 383.
2. C. M. King: *Control of stoats in the National Parks and Reserves of New Zealand.* National Parks
    Scientific Series, Department of Lands & Survey, Wellington, in press.
3. A. Moorehead: *Darwin and the Beagle*, Hamish Hamilton, London (1969), p.43.
4. C. M. King: Stoat observations. *Landscape* 12:12-15(1982).
5. Psalm 104:20-22, Authorized Version.
6. J. Kikkawa: Population distribution of land birds in temperate rainforest of southern New
    Zealand. *Transactions of the Royal Society of N.Z.* (*Zoology*) 7:215-277(1966).
7. C. M. King: Age structure and reproduction in feral New Zealand populations of the
    house mouse (*Mus musculus*), in relation to seedfall of southern beech. *N.Z. Journal of
    Zoology* 9:467-480 (1982).
8. *New Zealand Official Yearbook, 1982.* Government Printer, Wellington.
9. D. V. Merton: Controlling introduced predators and competitors on islands. In
    Endangered Birds, edited by S. A. Temple, Croom Helm, London (1978), p.125.
10. C. N. Challies: Effects of commercial hunting on red deer densities in the Arawhata
    Valley, South Westland, 1972-76. *N.Z. Journal of Forestry Science* 7:263-273(1977).
11. G. Caughley: *The Deer Wars.* Heinemann, Auckland (1983), Ch. 8.
12. P. J. Moors: Predation by mustelids and rodents on the eggs and chicks of native and
    introduced birds in Kowhai Bush, New Zealand. *Ibis* 125:137-154(1983).
13. E. B. Spurr: A theoretical assessment of the ability of bird species to recover from an
    imposed reduction in numbers, with particular reference to 1080 poisoning. *N.Z.
    Journal of Ecology* 2:46-63(1979).
14. A. J. Marshall, quoted on p.122 of W. D. L. Ride: A background to conservation in
    Australia. In *A Vanishing Heritage* (49th Congress of ANZAS), Nature Conservation
    Council, Wellington (1979), pp.113-138.
15. R. K. Murton: *Man and Birds.* Collins, London (1971), p.xv.
16. J. A. D. Flack and B. D. Lloyd: In *The ecology and control of rodents in New Zealand nature
    reserves*, edited by P. R. Dingwall, I. A. E. Atkinson and C. Hay. Departments of Lands &
    Survey, Information Series 4 (1978), pp.59-66.
    R. G. Powlesland: Comparison of time-budgets for mainland and Outer Chetwode Island
    populations of adult male South Island robins. *N.Z. Journal of Ecology* 4:98-105(1981).
    D. M. Hunt and B. J. Gill (eds): Ecology of Kowhai Bush, Kaikoura. *Mauri Ora* Special
    Publication No. 2 (1979), pp.1-54.

17. D. V. Merton: Cross-fostering of the Chatham Island black robin. *N.Z. Journal of Ecology* 6:156-157(1983).
    Dominion, 7 March 1984.
18. C. C. Ogle: Great Barrier Island Wildlife Survey. *Tane* 27:177-200(1981).
19. J. M. Diamond and C. R. Veitch: Extinctions and introductions in the New Zealand avifauna: cause and effect? *Science* 211:499-501(1981).
20. I. A. E. Atkinson: What's so special about Kapiti Island? *Forest and Bird* 212:12-15(1979).
    J. Kikkawa: A bird census on Kapiti Island. *Records of the Dominion Museum* 3:307-320(1960).
21. R. E. Brockie and A. Moeed: Animal biomass in native forest of the Orongorongo Valley, Wellington. *N.Z. Journal of Ecology* 4:131-132(1981).
22. T. Lovegrove: Unpublished report to the District Office of the Department of Lands & Survey, Wellington, July 1983.
23. C. R. Veitch: Eradication of cats from Little Barrier Island. *Landscape* 11:27-29(1982).
24. C. C. Ogle and J. Cheyne: *The wildlife and wildlife values of the Whangamarino wetlands.* N.Z. Wildlife Service, Fauna Survey Unit Report 28:1-76(1981).
25. H. Guthrie-Smith: *Tutira* (4th edition) A. H. & A. W. Reed, Wellington (1969), pp.215-6.
26. K. R. Hackwell and D. G. Dawson: Designing forest reserves. *Forest and Bird* 218:8-15(1980).
27. P. C. Bull, P. D. Gaze and C. J. R. Robertson: *Bird distribution in New Zealand: a provisional atlas 1969-1976.* Ornithological Society of New Zealand Inc., Wellington (1978).
28. I. G. Crook: Wildlife surveys and the future of North Westland forests. *N.Z. Wildlife Service Annual Report* 8 (1977), p.18.
29. C. Imboden: Conservation: A problem for the ecologist? *Forest and Bird* 211:2-6(1979).
30. C. A. Fleming: Scientific planning of reserves. *Forest and Bird* 196:15-18(1975).
31. M. E. Soulé and B. A. Wilcox: *Conservation Biology.* Sinauer Press, Mass. (1980). Ch. 9.
32. R. A. Falla, R. B. Sibson and E. G. Turbott: *The New Guide to the Birds of New Zealand.* Collins, Auckland (1981), p.73.
33. R. Pierce: The black stilt: endangered bird of the high country. *Forest and Bird* 217:15-18(1980).
34. J. T. Lokemoen, H. A. Doty, D. E. Sharp and J. E. Neaville: Electric fences to reduce mammalian predation on waterfowl nests. *Wildlife Society Bulletin* 10:318-323(1982).
35. R. Pierce: A record breeding season for the black stilts in the wild. *Forest and Bird* 228:30-31(1983).
36. Falla *et al. op.cit.,* p.87.
37. J. Kear and P. J. K. Burton: The food and feeding apparatus of the blue duck, *Hymenolaimus. Ibis* 113:483-493(1971).
38. P. R. Millener: *The Quaternary avifauna of the North Island, New Zealand.* PhD Thesis, University of Auckland (1981), p.592-593.
39. *Seminar on the takahe and its habitat*: Fiordland National Park Board, Invercargill (1978), p.32.
40. J. A. Mills and A. F. Mark: Food preferences of takahe in Fiordland National Park, New Zealand, and the effect of competition from introduced red deer. *Journal of Animal Ecology* 46:939-958(1977).
41. J. A. Mills, W. G. Lee, A. F. Mark and R. B. Lavers: Winter use by takahe (*Notornis mantelli*) of the summer-green fern (*Hypolepis millefolium*) in relation to its annual cycle of carbohydrates and minerals. *N.Z. Journal of Ecology* 3:131-137(1980).
42. *Seminar on the takahe and its habitat*, p.169.
43. *Ibid*, p.72.
44. J. A. Mills, *pers. comm.*

45. J. A. Mills, R. B. Lavers and M. C. Crawley: Takahe and the wapiti issue. *Forest and Bird* 225:2-5(1982).

46. N.Z. Wildlife Service data.

47. R. B. Lavers: Distribution of the North Island kokako (*Callaeas cinerea wilsoni*): a review. *Notornis* 25:165-185(1978).

48. R. Wilson: *From Manapouri to Aramoana – The battle for New Zealand's environment.* Earthworks Press, Auckland (1982), pp.120-123.
    L. W. Wright: Decision-making and the logging industry: an example from New Zealand. *Biological Conservation* 18:101-115(1980).

49. J. R. Leathwick, J. R. Hay and A. E. Fitzgerald: The influence of browsing by introduced mammals on the decline of North Island kokako. *N.Z. Journal of Ecology* 6:55-70 (1983).

50. J. R. Hay: The Kokako: Perspective and Prospect. *Forest and Bird* 15(1):6-11(1984).

51. J. G. Innes, *pers. comm.*

52. P. R. Millener, *op.cit.*, pp.613-616.

53. H. A. Best, *pers. comm.*

54. R. Powlesland, *pers. comm.*

55. P. J. Moors, *pers. comm.*

56. C. M. King: The biology of the stoat (*Mustela erminea*) in the National Parks of New Zealand. IV. Reproduction. *N.Z. Journal of Zoology* 9:103-118(1982).

57. B. D. Bell: The challenge of the 'stoat invasion' on Maud Island. *Forest and Bird* 227:12-14(1983).

58. J. A. Jackson: Alleviating problems of competition, predation, parasitism and disease in endangered birds. In *Endangered Birds*, edited by S. A. Temple. Croom Helm, London (1978), p.79.

59. M. C. Crawley: Wildlife Conservation in New Zealand. *N.Z. Journal of Ecology* 5:1-5 (1982).

60. G. M. Thomson: *The Naturalization of animals & plants in New Zealand.* Cambridge University Press, Cambridge (1922), p.82.

61. M. J. Daniel: Bionomics of the shiprat (*Rattus r. rattus*) in a New Zealand indigenous forest. *N.Z. Journal of Science* 15:313-341(1972).

62. P. R. Dingwall, I. A. E. Atkinson and C. Hay (eds): *The ecology and control of rodents in New Zealand nature reserves.* Department of Lands & Survey, Wellington (1978), p.175.

63. *Ibid*, p.138.

64. *Ibid*, pp.150, 183.

65. C. M. King: The reproductive tactics of the stoat (*Mustela erminea*) in New Zealand forests. In *Proceedings of the First Worldwide Furbearer Conference*, edited by J. A. Chapman and D. Pursely (1981), pp.443-468.

66. C. M. King and C. D. McMillan: Population structure and dispersal of peak-year cohorts of stoats (*Mustela erminea*) in two New Zealand forests, with especial reference to control. *N.Z. Journal of Ecology* 5:59-66(1982).

67. P. J. W. Langley and D. W. Yalden: The decline of the rarer carnivores in Great Britain during the nineteenth century. *Mammal Review* 7:95-116(1977).

68. R. K. Murton, *op.cit.*, Ch.2.

69. N. Livingston: The decline of our native birds. *New Zealand Wildlife* 39:47(1972).

70. W. D. L. Ride: A Background to conservation in Australia. In *A Vanishing Heritage*. Nature Conservation Council, Wellington (1979), p.116.

71. R. M. Lockley: Permits must not be granted to moor at the Snares Islands. *Forest and Bird* 228:2-5(1983).
    R. K. Dell, in *Proceedings of the Royal Society of N.Z.* 111:82-83(1983).

Notes to Chapter 6

(a) The Polynesian rat was not used for food in Hawaii, as it was in New Zealand – understandably so, because the Hawaiians had plenty of pork and chicken. Since the human colonists of both Hawaii and New Zealand came from the same place, one wonders why the Asian pig and fowl did not reach New Zealand.

(b) The excavations are still going on; much material remains to be identified and many islands have not been well searched, so Olson and James[8] emphasize that the results they have published so far are preliminary and incomplete.

(c) The figures quoted by Atkinson are different because he included subspecies, which Olson and James excluded. However, it does not much matter which system is used, so long as it is used consistently; the message is the same.

(d) The land areas (in sq. km) of the three groups span five orders of magnitude: Lord Howe $= 1.3 \times 10^1$; Hawaiian chain $= 1.6 \times 10^4$; New Zealand $= 2.6 \times 10^5$. Australia represents the next order up (7.7×10^6).

(e) Two thirds of Australia's land surface receives less than 500 mm of rain a year.

(f) The Aboriginals were resident at the same time that considerable climatic changes were transforming Australia at the end of the northern ice ages (felt in Australia as a cool, wet period). These changes must have affected the native fauna, and some species would probably have died out naturally, so it is even more difficult to distinguish the effects of prehistoric hunting than in New Zealand, where at least the human onslaught was delayed until long after the climatic adjustments were complete (p.48).

(g) The most recent survey of New Zealand's endangered birds found that the rate of extinction among the endemic species has been some eight times higher than that of the indigenous (native, but not endemic) species. Of all the species now listed as endangered or rare, 19% are endemic and 4% indigenous. Among the land and freshwater birds, 37% of the endemic species are extinct, endangered or rare, compared with only 6% of the indigenous species.[35] The link between extinction and endemism in New Zealand birds was made most explicit by R. M. McDowall, who showed a startlingly regular relationship between the degree of endemism of a group of birds, and the proportion of it that is now extinct.[36]

References to Chapter 6

1. S. J. Gould: *Ever since Darwin*. W. W. Norton, New York (1977), p.134.
2. S. J. Gould: *The Panda's Thumb*. W. W. Norton, New York (1980), p.289.
3. D. V. Merton: Controlling introduced predators and competitors on islands. In *Endangered Birds*, edited by S. A. Temple. Croom Helm, London (1978), p.121.
4. H. R. Recher, D. Lunney and I. Dunn: *A natural legacy. Ecology in Australia*, Pergamon Press (Australia) Pty Ltd (1979), p.78.
5. K. A. Hindwood: The birds of Lord Howe Island. *Emu* 40:1-86(1940).
6. *Ibid*, p.7.
7. *Ibid*, p.25.
8. S. L. Olson and H. F. James: Prodromus of the fossil avifauna of the Hawaiian Islands. *Smithsonian contributions to Zoology* 365:1-59(1982).
   S. L. Olson and H. F. James: Fossil birds from the Hawaiian Islands: Evidence for wholesale extinction by man before western contact. *Science* 217:633-635(1982).
9. S. Carlquist: *Hawaii. A Natural History* (2nd edition). Pacific Tropical Botanical Garden, Hawaii (1980).
10. P. T. Boag: More extinct birds. *Nature* 305:274-75(1983).
11. Olson and James, *op.cit.*, p.46.
12. P. V. Kirch: The impact of the pre-historic Polynesians on the Hawaiian ecosystem. *Pacific Science* 36:1-14(1982).

13. Olson & James, *op.cit.*, p.47.

14. P. Q. Tomich: *Mammals in Hawaii*. Bishop Museum Press, Honolulu (1969).

15. I. A. E. Atkinson: A reassessment of factors, particularly *Rattus rattus* L., that influenced the decline of endemic forest birds in the Hawaiian Islands. *Pacific Science* 31:109-33(1977).

16. P. Tomich, *op.cit.*, p.58.

17. I. Atkinson, *op.cit.*, p.126.

18. P. Boag, *op.cit.*, p.274.

19. I. Atkinson, *op.cit.*, p.116.

20. J. C. Greenway: *Extinct and vanishing birds of the world*. American Committee for International Wild Life Protection, New York (1958), pp.231-5.

    D. Day: *The doomsday book of animals*. Ebury Press, London (1981), p.92.

21. M. Hachisuka: *The dodo and kindred birds*. Witherby, London (1953).

22. E. Hyams: *The changing face of Britain*. Paladin Books, St Albans (1977).

23. M. Williamson: *Island populations*. Oxford University Press, Oxford (1981), p.198.

    G. B. Corbet and H. N. Southern: *The Handbook of British Mammals*. Blackwell Scientific Publications, Oxford (1977), pp.1-6.

24. R. K. Murton: *Man and Birds*. Collins, London (1971), p.103.

25. B. Campbell: Arctic invaders of the Highlands. *New Scientist* 63:507-9(1974).

    K. Williamson: New bird species admitted to the British and Irish lists since 1800. In *The changing fauna and flora of Britain*, edited by D. L. Hawksworth. Academic Press, London (1974), pp.221-27.

    R. K. Murton, *op.cit.*, Ch. 1.

26. G. B. Corbet: The distribution of mammals in historic times. In *The changing flora and fauna of Britain*, edited by D. L. Hawksworth. Academic Press, London (1974), pp.179-202.

27. Corbet and Southern, *op.cit.*, p.348.

28. P. R. F. Chanin and D. J. Jefferies: The decline of the otter *Lutra lutra* L. in Britain: an analysis of hunting records and discussion of causes. *Biological Journal of the Linnaean Society* 10:305-328(1978).

29. Recher *et al.*, *op.cit.*, pp.13-26.

30. A. Keast: Australian Mammals: Zoogeography and evolution. In *Evolution, Mammals and southern Continents*, edited by A. Keast *et al*. State University of New York Press, Albany (1972), pp.195-246.

31. Recher *et al.*, *op.cit.*, p.25.

    J. Flood: *Archaeology of the Dreamtime*. Collins, Sydney (1983), Ch. 12.

32. J. Robertson: *The Captain Cook Myth*. Angus & Robertson, Sydney (1981).

33. H. J. Frith: *Wildlife Conservation* (2nd edition). Angus & Robertson, Sydney (1979), p.9.

34. *Ibid*, p.115.

35. J. A. Mills and G. R. Williams: The status of endangered New Zealand birds. In *The Status of Endangered Australasian Wildlife*, edited by M. J. Tyler. Royal Zoological Society of South Australia (1979), p.153.

36. R. M. McDowall: Extinction and endemism in New Zealand birds. *Tuatara* 17:1-12(1969).

37. R. Henry: *The habits of the flightless birds of New Zealand*. Government Printer, Wellington (1903), p.19.

38. H. J. Frith, *op.cit.*, p.153.

    I. Linn and P. Chanin: More on the mink 'menace'. *New Scientist* 79:38-40(1978), with references to previous controversy.

39. S. Olson and H. James, *op.cit.*, p.48-49.

40. Recher *et al.*, *op.cit.*, p.191.

Notes to Chapter 7

(a) Past writers have put forward at least seven theories accounting for the known historic extinctions of New Zealand birds. Of the seven, the status of recent (post-Pleistocene) climatic change is disputed (p.48); orthogenetic trends, and the competition and diseases from exotic birds, seem unimportant except as minor contributory causes (pp.49, 80, 115). They are not dismissed, but the overwhelming weight of the evidence points to the other three: reduction in total forest area, dissection and degradation of the remaining forest, and the introduced predators, including human hunters.

(b) This is a false view which arises from comparing changes in the native fauna only since the arrival of the first European explorers. It leads to wrong ideas about the significance of that date, because it fosters a short-term view of extinctions which underestimates the importance of the changes that took place before there were Europeans to observe them.

(c) Any change in environmental conditions will eventually stimulate subtle genetic adaptations in the resident animals.[8] If the changes are not too great, this simply ensures that the animals remain well-adapted to their surroundings. This is a continuous process, and has been for aeons of time – as the Red Queen said, you have to run hard in order to stay in the same place. Only if the changes are so great or so rapid that they exceed the capacity of the species to respond does this constant running adjustment break down, and then species falter, stumble, and fall. We class them as 'endangered'; and then, all too often, as 'extinct'.

(d) The problem is not that the expeditions themselves waste resources, so long as they also do other valuable work in the process of checking up on the reported sighting, which of course they usually do.[13] The problem is that, *if* such birds were ever found still to exist, the long-term chances of so extremely endangered a species being rescued seem very small, when one considers that other species still in greater numbers (e.g. black stilts) are considered unlikely to last much longer (p.141). In the natural course of events, most extremely endangered species will become extinct, and can be prevented from doing so only by *preservation*, i.e. expensive, energy-intensive artificial management consciously intended to alter the natural course of evolution. When money is short, we may have to decide whether we want to invest in *conservation*, to protect wild nature as it is and as it really works, or to manage an artificial zoo.

(e) Obviously, there is nothing intrinsically wrong in operating a business for profit, provided it can be done without causing long-term damage to other people's interests. I applaud the efficient harvesting of the existing pine plantations; but to continue to fell native forest now, in order to extend the domain of the pines, or for any other reason, is indefensible.

(f) Appeals for conservation on the grounds of some supposed right to exist, even from eminent persons like HRH Prince Philip, are well-intentioned but are bad biology, and do no service to the development of rational conservation ethics. For a better effort, see Ratcliffe.[17] My own view is simply that careless use by the present generation of the natural resources of the earth amounts to theft from the next generation. To destroy what is not one's own is recognized as wrong in every civilised moral code.

References to Chapter 7

1. P. M. Coker and C. Imboden: *Wildlife values and wildlife conservation in South Westland*. N.Z. Wildlife Service, Fauna Survey Unit Report 21:1-84(1980).
2. New Zealand Ecological Society: The future of west coast forests and forest industries. *N.Z. Journal of Ecology* 1:166-172(1978).
3. N. Myers: *The Sinking Ark*. Pergamon Press, Oxford (1979).
4. C. A. Fleming: The extinction of moas and other animals during the Holocene period. *Notornis* 10:113-117(1962).

5. C. M. King: *Control of stoats in the National Parks and Reserves of New Zealand*. National Parks Scientific Series, Department of Lands & Survey, Wellington (in press).

6. N. Myers, *op.cit.,* p.43.

7. O. H. Frankel and M. Soulé: *Conservation and evolution*. Cambridge University Press, Cambridge (1981), p.4.

8. R. J. Berry: Conservation aspects of the genetical constitution of populations. In *Symposia of the British Ecological Society* 11:177-206(1971).

9. C. M. King and J. E. Moody: The biology of the stoat (*Mustela erminea*) in the National Parks of New Zealand. III. Morphometic variation in relation to growth, geographical distribution, and colonization. *N.Z. Journal of Zoology* 9:81-102(1982), and references cited.

10. Paraphrased from Myers, *op. cit.*, p.X.

11. P. F. Jenkins: Management of threatened New Zealand species. In *A vanishing heritage*. Nature Conservation Council, Wellington (1979), pp.178-185.

12. L. W. Wright: Decision-making and the logging industry: an example from New Zealand. *Biological Conservation* 18:101-115 (1980).

13. M. N. Clout and J. R. Hay: South Island kokako (*Callaeas cinerea cinerea*) in *Nothofagus* forest. *Notornis* 28:256-259 (1981).

14. M. C. Crawley: Wildlife conservation in New Zealand (Presidential Address). *N.Z. Journal of Ecology* 5:1-5 (1982).

15. N. Myers, *op. cit.*, p.51.

16. A. Esler: What is left to conserve? *Landscape* 6:24-30 (1979).

17. D. A. Ratcliffe: Thoughts towards a philosophy of nature conservation. *Biological Conservation* 9:45-53 (1976).

18. R. K. Murton: *Man and Birds*. Collins, London (1971), p.XIV.

19. H. Martin: The protection of native birds. *Transactions and Proceedings of the N.Z. Institute* 18:112-117 (1886).

TABLES

Table 1: The broad vegetation cover of New Zealand in pre-Polynesian, early European and modern times

	Pre-Polynesian Area[1]	%[2]	1840 Area	%	Present Area	%
Native forest and light woodland	c.21,102	78	c.14,000	53	6,246	23
Alpine zone	3,725	14	3,725	14	3,725	14
Open country[3]						
'Arid zone' (grassland & scrub)	c.1,225	5 ⎫	7,744[5]	29 ⎫		
Induced fern and tussock	–	– ⎭		⎬	13,940	52
'Improved' and 'Other' grazing land	–	–	–	– ⎭		
Lakes[4]	324	1	324	1	340	1
Riverbeds	150	0.6	150	0.6	160	0.6
Swamps	c.230	0.9	455	2 ⎫		
Dunes	c.60	0.2	130	0.5 ⎭	882[5]	3
Exotic forests	–	–	–	–	740	3
Cropping land	–	–	–	–	429	2
Towns and cities	–	–	–	–	370	1
Total	26,832[4]		26,528[2]		26,832	

[1] In thousands of hectares.

[2] Percentage of the land area of New Zealand, including all the offshore but not the subantarctic islands. The figures for 1840 are calculated for the area of the North and South Islands only, but the error in the percentages thus introduced is probably immaterial in view of the probable errors in compiling the figures for that time, and also for pre-Polynesian times. Because accurate measures are now impossible, these data are acknowledged to be 'best estimates' only (G. C. Kelly, *pers. comm.*).

[3] Different definitions in each time period unavoidable. Later categories include earlier ones.

[4] Natural lakes only. I have subtracted 16,000 ha for artificial (hydro-electric) lakes, so the total for the first column does not add up to 26,832 (*cf.* Kelly, 1980).

[5] By subtraction (includes other habitats classified under 'miscellaneous').

Sources:

G. C. Kelly, in *Land alone endures*, edited by L. F. Molloy *et al.* DSIR Discussion Paper 3:63-87(1980).

M. S. McGlone: Polynesian deforestation of New Zealand: a preliminary synthesis. *Archaeology in Oceania* 18:11-25(1983).

I. Wards: *The New Zealand Atlas*. Government Printer, Wellington (1976). The vegetation of 1840 is mapped on pp.104-105; areas calculated by G. C. Kelly and C. M. King.

Table 2: The historic declines or extinctions of endemic New Zealand mainland birds, and the predators that could have affected them (For scientific names see Appendix)

Species of wild or feral predators, and the period in which they were most active	Birds of the North and South Islands suffering decline or extinction during the same period[1] (some survive on offshore islands – See Table 7)
a) *Polynesian period (c. 750-1769)*	
man (hunters)	12 species of moa; 7 of large waterfowl;
kiore	5 of rails; 3 of eagles and hawks; one
Maori dog	pelican; one coot; one crow; one
	owlet-nightjar; one merganser; one
	snipe; and at least two wrens
	takahe
	kakapo ⎫ on North Island
	little spotted kiwi ⎭
b) *Early European period (1769-1884)*	
man (explorers and collectors)	New Zealand quail in both main islands
kiore	North Island laughing owl
feral Maori × European dog	stitchbird
Norway rat	North Island saddleback
feral cat	North Island thrush
ship rat (only in NI, after 1860s)	huia
	North Island bush wren
c) *Later European period (1884-1984)*	
Norway rat	North Island kokako
ship rat	South Island bush wren
feral cat	South Island thrush
ferret (from 1882)	Eastern or Buff weka
stoat	South Island laughing owl
weasel	South Island saddleback
hedgehog	South Island kokako
	little spotted kiwi in South Island
	orange-fronted parakeet
	black stilt
	brown teal
	blue duck
	red-crowned parakeet

[1] Excludes species that declined temporarily and later recovered. Note that, because birds listed as affected mainly in the later periods were probably already under stress from changes taking place in earlier periods, this list tends to exaggerate the effects of predators that arrived only in the third period, such as stoats. However, there were areas, such as the south and west of the South Island, which were still rich in birdlife when stoats arrived, along with shiprats, in the 1890s.

Sources:

P. R. Millener: *The Quaternary avifauna of the North Island, New Zealand*. PhD Thesis, Auckland University (1981). Also in *Notornis* 29: 165-170 (1982).

J. A. Mills and G. R. Williams: in *The Status of Endangered Australasian Wildlife*, edited by M. J. Tyler. Royal Zoological Society of South Australia (1979), pp.147-168.

G. R. Williams and D. R. Given (editors): *The Red Data Book of New Zealand*. Nature Conservation Council, Wellington (1981).

Table 3: Actual and expected[1] total losses of land and freshwater native bird species in the North Island[2], in relation to forest clearance

	Polynesian era (1000 to c.1800 AD)	European era (c. 1800 to present)	Total (1000 to present)
Area of native forest (millions of hectares) present at beginning and end of period	10.9-8.4 mh	8.4-2.5 mh	10.9-2.5 mh
% reduction in forest by the end of the period	23% of 10.9	70% of 8.4	77% of 10.9
Total number of land and freshwater native bird species, present at beginning and end of period	73-52	52-39	73-39
Full species becoming extinct in the North Island during this period (including those surviving on the South Island and on offshore islands)	21	13	34
% reduction in native avifauna by end of the period	29% of 73	25% of 52	46% of 73
Expected number of extinctions[3] due to forest reduction alone	about 3%	about 16%	about 21%

[1] Using the 'rule of thumb' that 10% of the resident birds will be lost for every 50% loss of forest (p.54).

[2] No comparable palaeontological analysis for the South Island exists.

[3] The expected losses in the bottom line were estimated by plotting the theoretical losses of birds in 10% steps against loss of forest in 50% steps.

From 100% of both at the beginning, the curve declines slowly for a long time: when only 20% of forest is left, there is still about 77% of the original birds. Below 10% of forest left, the curve for birds suddenly plunges, and can no longer be plotted. It does not necessarily follow that all native forest birds will disappear if all the forest goes (some could survive in farmlands, gardens and pine plantations).

Sources:

P. R. Millener, G. C. Kelly, *op.cit.* (see Tables 1 and 2).

J. L. Nicholls: The past and present extent of New Zealand's indigenous forests. *Environmental conservation* 7:309-310(1980), and *pers. comm.*

Table 4: Summary distribution of the 153 separate populations of birds known to have become extinct in New Zealand since 1000 AD (excluding the outlying islands) according to whether or not they could have been affected by stoats

	Island groups[1]				
	North	*South*	*Stewart*	*Chatham*	*Total*
Never in contact with stoats:[2]					
Extinct	46	41	14	29	130
Rare/Endangered	2	0	1	4	7
Possibly affected by stoats:[3]					
Extinct	1	4	–	–	5
Rare/Endangered	1	6	–	–	11[4]
				Total	153[5]

[1] Principal and inshore islands in each area.

[2] Disappeared from the North or South Islands before 1884, or from Stewart, Chatham or offshore islands which stoats never reached.

[3] Disappeared from the North or South Islands since 1884.

[4] Plus 4 species common to both North and South Island.

[5] A conservative estimate.

Source:

C. M. King: *The control of stoats in the National Parks and Reserves of New Zealand.* National Parks Scientific Series, Department of Lands and Survey, Wellington (in press).

Table 5: Presumed causes of extinction of birds on islands throughout the world since 1600AD

	Indian Ocean	*Atlantic Ocean*	*Pacific Ocean*	*Totals*[1]
Predation	0	16	96	112[2]
Hunting	7	13	20	40
Habitat destruction	1	5	46	52
Competition	0	0	18	18
Disease	0	0	15	15
Genetic swamping	0	0	2	2
Weather	0	1	0	1
Unknown	13	4	11	28

[1] The grand total here is 268, whereas the total number of species and subspecies known to have become extinct on islands is about 163 (the source author, W. B. King, merely says '93% of 92 species and 83 subspecies'). The causes listed are therefore contributory, not mutually exclusive, and cannot be added.

[2] King states that 'Predation has been implicated as a cause in 70% of extinctions of birds on islands. Of these, 54% are attributed to rats, among which the black [=ship] rat *Rattus rattus* is the most serious predator, although the Norway rat *R. norvegicus* and the Polynesian rat *R. exulans* [kiore] are believed also to have made contributions . . . Cats are implicated in 26% of extinctions by predation, while . . . the mustelids *Mustela nivalis* and *M. erminea* have been implicated once or several times each.'

In 1982, the International Council for Bird Preservation organized a symposium on the management of island avifaunas. In his introductory review paper 'Island birds: will the future repeat the past?', King remarked that the contemporary endangered island birds are clearly destined to become tomorrow's extinctions, and that 'island groups such as the Hawaiian Islands, the Mascarenes, New Zealand and its outliers, and the Seychelles, where extinction has already decimated species diversity, will continue to be the largest source of extinctions in the future'. However, he also added that 'nowhere have the techniques involved in island rehabilitation been put to better use than in New Zealand'.

Source:

W. B. King: Ecological basis of extinction in birds. *Proceedings of the 17th International Congress of Ornithology* (1980), p.905-911; and Proceedings of the ICBP Conference, Cambridge, August 1982, in press.

Table 6: List of additions to the New Zealand avifauna, self-introduced since 1840
(with date of first known breeding)

N.Z. Checklist number	Species
16.1	hoary-headed grebe (1976)
17	Australian little grebe (1977)
90	whitefaced heron (1941)
101	royal spoonbill (1950s)
110	grey teal (? 1930s)
141	Australian coot (1950s)
145	spur-winged plover (1932)
154	black-fronted dotterel (1950s)
245	welcome swallow (1950s)
263	silvereye (1850s)
	Total: 10

Sources:

F. C. Kinsky: *Annotated checklist of the birds of New Zealand*, Ornithological Society of New Zealand Inc. (1970), and the *Amendments and additions to the 1970 checklist*, published in the supplement to *Notornis* 27 (1980).

Table 7: Could control of predators possibly prevent any further extinctions of New Zealand birds?

Threatened species that can only or best be preserved on islands free of most predators[1]	Threatened species that can only or best be preserved on the mainland
stitchbird (Little Barrier Is.)	takahe
N. I. saddleback ⎱ several islands S. I. saddleback ⎰	black stilt
black robin (Mangere Is.)	N. I. kokako[2]
little spotted kiwi (Kapiti Is.)	kakapo (Stewart Is.)
buff weka (Chatham Is.)	
brown teal (Great Barrier Is.)	blue duck
red-crowned parakeet (numerous islands)	
Chatham Island pigeon	
Chatham Island snipe	
Chatham Island oystercatcher	
Chatham Island petrel	
Chatham Island taiko	
Forbes parakeet (Chatham Is.)	
shore plover (South-East Is.)	
orange-fronted parakeet	
Kermadec parakeet	
Kermadec storm petrel	
codfish fernbird	
Auckland Island rail	
Auckland Island snipe	
Auckland Island dotterel	
Total: 22	Total: 5

[1] Not all these islands are entirely predator-free, but most are free enough of at least of the worst predators (cats, ship rats, and mustelids) to give these bird populations a good chance of survival.

[2] If the N. I. kokako settles down on Little Barrier Island, its chance of survival there might be better than in a sufficiently large, legally secure mainland forest protected by any practicable programme of predator control. However, both policies contain the risk of failure, for different reasons (failure of the transferred birds to establish a viable population on the island, or failure of predator control to give adequate protection on the mainland).

APPENDIX
SCIENTIFIC NAMES OF ANIMALS AND PLANTS DISCUSSED UNDER THEIR COMMON NAMES IN THE TEXT

Bat, Steward Island short-tailed, *Mystacina tuberculata*
Beech, Southern, *Nothofagus* sp.
       Red, *N. fusca*
Bellbird, *Anthornis melanura*
Blackbird, European, *Turdus merula*
        Vinous tinted, *T. xanthopus vinitinctus*
Bracken, *Pteridium aquilinum*
Cabbage tree, *Cordyline australis*
Cat, *Felis catus*
Cawcaw, see Kaka
Chaffinch, *Fringilla coelebs*
Chamois, *Rupicapra rupicapra*
Coot, Australian, *Fulica atra australis*
     Extinct, *F. chathamensis*
Crake, Marsh, *Porzana pusilla affinis*
      Spotless, *P. tabuensis plumbea*
Creeper, Brown, *Finschia novaeseelandiae*
Crow, Extinct, *Palaeocorax moriorum*
Deer, Fallow, *Dama dama*
     Red, *Cervus elaphus*
     Sambar, *C. unicolor*
     Wapiti, *C. canadensis*
Dingo, see Dog
Dog, domestic, *Canis familiaris*
Dotterell, Auckland Island, *Charadrius bicinctus exilis*
      New Zealand, *C. obscurus*
      Black-fronted, *C. melanops*
Duck, Blue, *Hymenolaimus malacorhynchos*
     Paradise, *Tadorna variegata*
Dunnock, *Prunella modularis*
Eagle, Extinct Chatham Island, *Ichthyophaga australis*
Falcon, *Falco novaeseelandiae*
Fantail, New Zealand, *Rhipidura fuliginosa*
     Lord Howe Island, *R. cervina*
Fern, *aruhe*, see Bracken
Fernbird, Chatham Island, *Bowdleria punctata rufescens*
      Codfish Island, *B. p. wilsoni*
      North Island, *B. p. vealeae*
      South Island, *B. p. punctata*
      Stewart Island, *B. p. stewartiana*
Ferret, *Mustela furo*
Flax, *Phormium tenax*
Flyeater, Lord Howe Island, *Royigerygone insularis*

Fuchsia, *Fuchsia* sp.
Gallinule, white, *Porphyrio albus albus*
Goat, *Capra hircus*
Goose, extinct, *Cnemiornis* sp.
Grebe, Australian little, *Tachybaptus novaehollandiae*
        Crested, *Podiceps cristatus*
        Hoary-headed, *P. poliocephalus*
Gull, Black billed, *Larus bulleri*
Hare, *Lepus capensis*
Harrier, Australasian, *Circus approximans gouldi*
        Extinct, *C. eylesi*
Hedgehog, *Erinaceus europaeus*
Heron, whitefaced, *Ardea novaehollandiae*
Huia, *Heteralocha acutirostris*
Kaka, *Nestor meridionalis*
Kakapo, *Strigops habroptilus*
Kamahi, *Weinmannia racemosa*
Kauri, *Agathis australis*
Kea, *Nestor notabilis*
Kingfisher, *Halcyon sancta*
Kiore, see Rat, Polynesian
Kiwi, Little Spotted, *Apteryx owenii*
      Brown, *A. australis*
Kokako, North Island, *Callaeas cinerea wilsoni*
        South Island, *C. c. cinerea*
Kuri, see Dog
Magpie, Lord Howe Island, *Strepera graculina crissalis*
Manuka, *Leptospermum scoparium*
Matagouri, *Discaria toumatou*
Merganser, *Mergus australis*
Mongoose, Small Indian, *Herpestes auropunctatus*
Morepork, Ninox n. novaeseelandiae
Mouse, *Mus musculus*
Owl, Laughing, North Island, *Sceloglaux albifacies rufifacies*
      Laughing, South Island, *S. a. albifacies*
      Lord Howe Island, *Spilaglaux novaeseelandiae albaria*
      Short-eared, *Asio flammeus*
Owlet-nightjar, Extinct, *Megaegotheles novaezealandiae*
Oystercatcher, Chatham Island, *Haematopus chathamensis*
Parakeet, Forbes, *Cyanoramphus auriceps forbesi*
        Kermadec, *C. novaezelandiae cyanurus*
        Lord Howe Island, *C. n. subflavescens*
        Macquarie, *C. n. erythrotis*
        Orange-fronted, *C. malherbi*
        Red-crowned, *C. novaezelandiae*
        Yellow-crowned, *C. auriceps auriceps*
Parrot, Night, *Geopsittacus occidentalis*
Pelican, Extinct, *Pelecanus conspicillatus novaeseelandiae*
Petrel, Chatham Island, *Pterodroma axillaris*
      Kermadec Storm, *Pelagodroma marina albiclunis*

Pig, *Sus scrofa*
Pigeon, Chatham Island, *Hemiphaga novaeseelandiae chathamensis*
      Lord Howe Island, *Janthoenas godmanae*
      New Zealand, *H. n. novaeseelandiae*
Piopio, North Island, *Turnagra capensis tanagra*
      South Island, *T. c. capensis*
Pipit, New Zealand, *Anthus novaeseelandiae*
Plover, Shore, *Thinornis novaeseelandiae*
      Spur-winged, *Lobibyx novaehollandiae*
Possum, Brush-tailed, *Trichosurus vulpecula*
Pukeko, *Porphyrio porphyrio melanotis*
Quail, Australian stubble, *Coturnix novaezealandiae pectoralis*
      New Zealand, *C. n. novaezealandiae*
Rabbit, *Oryctolagus cuniculus*
Rail, Auckland Island, *Rallus pectoralis muelleri*
      Banded, *Rallus philippensis assimilis*
      Laysan Island, *Porzana palmeri*
Rat, Norway, *Rattus norvegicus*
      Polynesian, *R. exulans*
      Ship, *R. rattus*
Rata, *Metrosideros* sp.
Redpoll, *Acantha flammea*
Robin, Black, *Petroica traversi*
      North Island, *P. australis longipes*
      South Island, *P. a. australis*
      Stewart Island, *P. a. rakiura*
Rifleman, *Acanthisitta chloris*
Saddleback, North Island, *Philesturnus carunculatus rufusater*
      South Island, *P. c. carunculatus*
Scaup, *Aythya novaeseelandiae*
Seal, Fur, *Arctocephalus forsteri*
      Leopard, *Hydrurga leptonyx*
      Southern elephant, *Mirounga leonina*
Sealion, Hooker's, *Phocarctos hookeri*
Shag, White throated, *Palacrocorax melanoleucos*
Shoveler, *Anas rhynchotis variegata*
Silvereye, New Zealand, *Zosterops lateralis*
      Lord Howe Island, *Nesozosterops strenua*
Skylark, *Alauda arvensis*
Snipe, Auckland Island, *Coenocorypha a. aucklandica*
      Chatham Island, *C. a. pusilla*
      New Zealand, *Coenocorypha* sp.
      Stewart Island, *C. a. iredalei*
Spaniard, *Aciphylla* sp.
Sparrow, *Passer domesticus*
Spoonbill, Royal, *Platalea leucorodia regia*
Starling, European, *Sturnus vulgaris*
      Lord Howe Island, *Aplonis fusca hullianus*
Stilt, Black, *Himantopus novaezealandiae*
      Pied, *H. himantopus leucocephalus*

Stitchbird, *Notiomystis cincta*
Stoat, *Mustela erminea*
Swallow, Welcome, *Hirundo tahitica*
Takahe, *Nortornis mantelli*
Taiko, Chatham Island, *Pterodroma magentae*
Teal, Brown, *Anas aucklandica chlorotis*
      Grey, *Anas gibberifrons gracilis*
Tern, Black-fronted, *Sterna albostriata*
Thar, Himalayan, *Hemitragus jemlahicus*
Thrush, Song, *Turdus philomelos*
       Native, see Piopio
Tit, Pied, *Petroica macrocephala toitoi*
    Yellow-brested, *P. m. macrocephala*
    Chatham Island, *P. m. chathamensis*
Totara, Hall's, *Podocarpus hallii*
Tuatara, *Sphenodon punctatus*
Tui, *Prosthemadera novaeseelandiae*
Warbler, Chatham Island, *Gerygone albofrontata*
      Grey, *G. igata*
Wapiti, see Deer
Weasel, *Mustela nivalis*
Weka, *Gallirallus australis*
      Buff, *G. a. hectori*
Whiteye, see Silvereye
Whitehead, *Mohoua albicilla*
Wild Irishman, see Matagouri
Woodhen, Lord Howe Island, *Tricholiminas sylvestris*
Wren, Bush, North Island, *Xenicus longipes stokesi*
        South Island, *X. l. longipes*
        Rock, *X. gilviventris*
        Stead's, *X. l. variabilis*
        Stephens Island, *X. lyalli*
Yellowhead *Mohoua ochrocephala*

INDEX

Entries in *italic* refer to illustrations.

DATE DUE

GAYLORD   PRINTED IN U.S.A

DEMCO

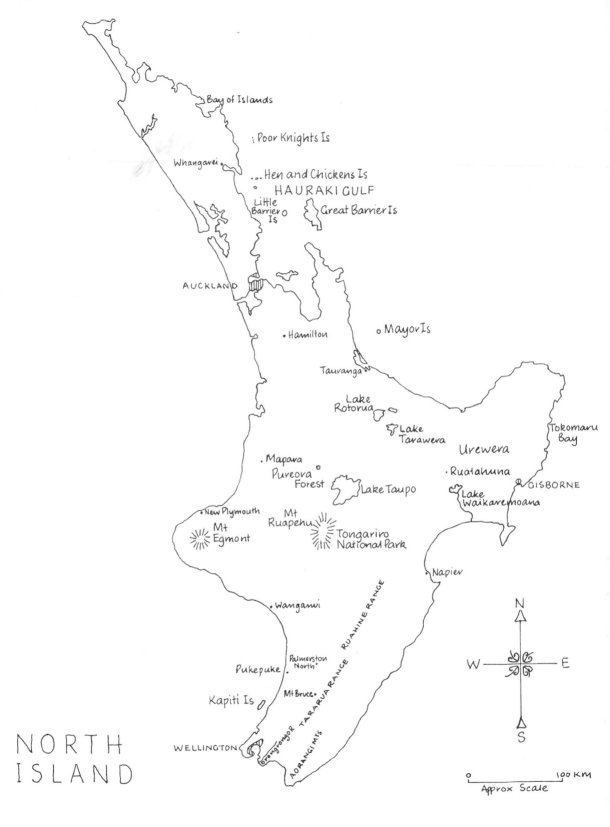

Three Kings Is

Bay of Islands

Poor Knights Is

Whangarei

Hen and Chickens Is

HAURAKI GULF

Little
Barrier
Is

Great Barrier Is

AUCKLAND

Hamilton

Mayor Is

Tauranga

Lake
Rotorua

Lake
Tarawera

Tokomaru
Bay

Urewera

Mapara

Pureora
Forest

Ruatahuna

Lake Taupo

Lake
Waikaremoana

GISBORNE

New Plymouth

Mt
Ruapehu

Mt
Egmont

Tongariro
National Park

Napier

Wanganui

RUAHINE RANGE

TARARUA RANGE

Pukepuke

Palmerston
North

Kapiti Is

Mt Bruce

Orongorongo

AORANGI MTS

WELLINGTON

N

W — E

S

NORTH
ISLAND

0   100 KM

Approx Scale